ASE Test Preparation Series

Automobile Test

Service Consultant (Test C1)

4th Edition

THOMSON
DELMAR LEARNING™

Australia Canada Mexico Singapore Spain United Kingdom United States

Thomson Delmar Learning's ASE Test Preparation Series

Automobile Test for Service Consultant (Test C1), 4th Edition

Vice President, Technology Professional Business Unit:
Gregory L. Clayton

Product Development Manager:
Kristen Davis

Product Manager:
Kim Blakey

Editorial Assistant:
Vanessa Carlson

Director of Marketing:
Beth A. Lutz

Marketing Specialist:
Brian McGrath

Marketing Coordinator:
Marissa Maiella

Production Manager:
Andrew Crouth

Production Editor:
Kara A. DiCaterino

Senior Project Editor:
Christopher Chien

XML Architect:
Jean Kaplansky

Cover Design:
Michael Egan

Cover Images:
Portion courtesy of DaimlerChrysler Corporation

Printed in the United States of America.
1 2 3 4 5 XXX 10 09 08 07 06

For more information contact:
Thomson Delmar Learning
Executive Woods
5 Maxwell Drive, PO Box 8007,
Clifton Park, NY 12065-8007
Or find us on the
World Wide Web
at www.delmarlearning.com
or www.trainingbay.com

ISBN: 1-4180-3889-X

NOTICE TO THE READER

Publisher does not warrant or guarantee any of the products described herein or perform any independent analysis in connection with any of the product information contained herein. Publisher does not assume, and expressly disclaims, any obligation to obtain and include information other than that provided to it by the manufacturer.

The reader is expressly warned to consider and adopt all safety precautions that might be indicated by the activities herein and to avoid all potential hazards. By following the instructions contained herein, the reader willingly assumes all risks in connection with such instructions.

The publisher makes no representation or warranties of any kind, including but not limited to, the warranties of fitness for particular purpose or merchantability, nor are any such representations implied with respect to the material set forth herein, and the publisher takes no responsibility with respect to such material. The publisher shall not be liable for any special, consequential, or exemplary damages resulting, in whole or part, from the readers' use of, or reliance upon, this material.

Contents

Section 5 Sample Test for Practice

Section 6 Additional Test Questions for Practice

Section 7 Appendices

Preface

Delmar Learning is very pleased that you have chosen our ASE Test Preparation Series to prepare yourself for the automotive ASE Examination. These guides are available for all of the automotive areas including A1–A8, the L1 Advanced Diagnostic Certification, the P2 Parts Specialist, the C1 Service Consultant and the X1 Undercar Specialist. These guides are designed to introduce you to the Task List for the test you are preparing to take, give you an understanding of what you are expected to be able to do in each task, and take you through sample test questions formatted in the same way the ASE tests are structured.

If you have a basic working knowledge of the discipline you are testing for, you will find Delmar Learning's ASE Test Preparation Series to be an excellent way to understand the "must know" items to pass the test. These books are not textbooks. Their objective is to prepare the technician who has the requisite experience and schooling to challenge ASE testing. It cannot replace the hands-on experience or the theoretical knowledge required by ASE to master vehicle repair technology. If you are unable to understand more than a few of the questions and their explanations in this book, it could be that you require either more shop-floor experience or further study. Some resources that can assist you with further study are listed on the rear cover of this book.

Each book begins with an item-by-item overview of the ASE Task List with explanations of the minimum knowledge you must possess to answer questions related to the task. Following that there are 2 sets of sample questions followed by an answer key to each test and an explanation of the answers to each question. A few of the questions are not strictly ASE format but were included because they help teach a critical concept that will appear on the test. We suggest that you read the complete Task List Overview before taking the first sample test. After taking the first test, score yourself and read the explanation to any questions that you were not sure about, including the questions you answered correctly. Each test question has a reference back to the related task or tasks that it covers. This will help you to go back and read over any area of the task list that you are having trouble with. Once you are satisfied that you have all of your questions answered from the first sample test, take the additional tests and check them. If you pass these tests, you will be prepared to do well on the ASE test.

Our Commitment to Excellence

The 4th edition of Delmar Learning's ASE Test Preparation Series has been through a major revision with extensive updates to the ASE's task lists, test questions, and answers and explanations. Delmar Learning has sought out the best technicians in the country to help with the updating and revision of each of the books in the series.

About the Series Advisor

To promote consistency throughout the series, a series advisor took on the task of reading, editing, and helping each of our experts give each book the highest level of accuracy possible. Dan Perrin has served in the role of Series Advisor for the 4th edition of the ASE Test Preparation Series. Dan began ASE testing with the first series of tests in 1972 and has been continually certified ever since. He holds ASE master status in automotive, truck, collision, and machinist. He is also L1, L2, and alternated fuels certified, along with some others that have expired. He has been an automotive educator since 1979, having taught at the secondary, post-secondary, and industry levels. His service includes participation on boards that include the North American Council of Automotive Teachers (NACAT), the Automotive Industry Planning Council (AIPC), and the National Automotive Technicians Education Foundation (NATEF). Dan currently serves as the Executive Manager of NACAT and Director of the NACAT Education Foundation.

Thanks for choosing Delmar Learning's ASE Test Preparation Series. All of the writers, editors, Delmar Staff, and myself have worked very hard to make this series second to none. I know you are going to find this ook accurate and easy to work with. It is our o ective to constantly improve our product at Delmar y responding to feed ack.

If you have any uestions concerning the ooks in this series, you can email me at autoexpert@trainingbay.com.

Dan Perrin
Series Advisor

Preface

Our Commitment to Excellence

About the Series Advisor

1 The History and Purpose of ASE

ASE began as the National Institute for Automotive Service Excellence (NIASE). It was founded as a non-profit independent entity in 1972 by a group of industry leaders with the single goal of providing a means for consumers to distinguish between incompetent and competent technicians. It accomplishes this goal by testing and certification of repair and service professionals. From this beginning it has evolved to be known simply as ASE (Automotive Service Excellence) and today offers more than 40 certification exams in automotive, medium/heavy duty truck, collision, engine machinist, school bus, parts specialist, automobile service consultant, and other industry-related areas. At this time there are more than 400,000 professionals with current ASE certifications. These professionals are employed by new car and truck dealerships, independent garages, fleets, service stations, franchised service facilities, and more. ASE continues its mission by also providing information that helps consumers identify repair facilities that employ certified professionals through its Blue Seal of Excellence Recognition Program. Shops that have a minimum of 75% of their repair technicians ASE certified and meet other criteria can apply for and receive the Blue Seal of Excellence Recognition from ASE.

ASE recognized that educational programs serving the service and repair industry also needed a way to be recognized as having the faculty, facilities, and equipment to provide a quality education to students wanting to become service professionals. Through the combined efforts of ASE, industry, and education leaders, the non-profit National Automotive Technicians Education Foundation (NATEF) was created to evaluate and recognize training programs. Today more than 2000 programs are ASE certified under the standards set by the service industry. ASE/NATEF also has a certification of industry (factory) training program known as CASE. CASE stands for Continuing Automotive Service Education and recognizes training provided by replacement parts manufacturers as well as vehicle manufacturers.

ASE certification testing is administered by the American College Testing (ACT). Strict standards of security and supervision at the test centers insure that the technician who holds the certification earned it. Additionally ASE certification also requires that the person passing the test to be able to demonstrate that they have two years of work experience in the field before they can be certified. Test questions are developed by industry experts that are actually working in the field being tested. There is more detail on how the test is developed and administered in the next section. Paper and pencil tests are administered twice a year at over seven hundred locations in the United States. Computer based testing is now also available with the benefit of instant test results at certain established test centers. The certification is valid for five years and can be recertified by retesting. So that consumers can recognize certified technicians, ASE issues a jacket patch, certificate, and wallet card to certified technicians and makes signs available to facilities that employ ASE certified technicians.

You can contact ASE at any of the following:

National Institute for Automotive Service Excellence
101 Blue Seal Drive S.E.
Suite 101
Leesburg, VA 20175
Telephone 703-669-6600
FAX 703-669-6123
www.ase.com

2 Take and Pass Every ASE Test

Participating in an Automotive Service Excellence (ASE) voluntary certification program gives you a chance to show your customers that you have the "know-how" needed to work on today's modern vehicles. The ASE certification tests allow you to compare your skills and knowledge to the automotive service industry's standards for each specialty area.

If you are the "average" automotive technician taking this test, you are in your mid-thirties and have not attended school for about fifteen years. That means you probably have not taken a test in many years. Some of you, on the other hand, have attended college or taken postsecondary education courses and may be more familiar with taking tests and with test-taking strategies. There is, however, a difference in the ASE test you are preparing to take and the educational tests you may be accustomed to.

How are the tests administered?

ASE test are administered at over 750 test sites in local communities. Paper and pencil tests are the type most widely available to technicians. Each tester is given a booklet containing questions with charts and diagrams where required. You can mark in this test booklet but no information entered in the booklet is scored. Answers are recorded on a separate answer sheet. You will enter your answers, using a number 2 pencil only. ASE recommends you bring four sharpened number 2 pencils that have erasers. Answer choices are recorded by coloring in the blocks on the answer sheet. The answer sheets are scanned electronically and the answers tabulated. For test security, test booklets include randomly generated questions. Your answer key must be matched to the proper booklet so it is important to correctly enter the booklet serial number on the answer sheet. All instructions are printed on the test materials and should be followed carefully.

ASE has introduced Computer Based Testing (CBT) at some locations. While the test content is the same for both testing methods the CBT tests have some unique requirements and advantages. It is strongly recommended that technicians considering the CBT tests go the ASE web page at www.ASE.com and review the conditions and requirements for this type of test. There is a demonstration of a CBT that allows you to experience this type of test before you register. Some technicians find this style of testing provides an advantage, while others find operating the computer a distraction. One significant benefit of CBT is the availability of instant results. You can receive your test results before you leave the test center. CBT testing also offers increased flexibility in scheduling. The cost for taking CBTs is slightly higher than paper and pencil tests and the number of testing sites is limited. The first time test taker may be more comfortable with the paper and pencil tests but technicians now have a choice.

Who Writes the Questions?

The questions are written by service industry experts in the area being tested. Each area will have its own technical experts. Questions are entirely job related. They are designed to test the skills you need to be a successful technician. Theoretical knowledge is important and necessary to answer the questions, but the ability to apply that knowledge is the basis of ASE test questions.

Each question has its roots in an ASE "item-writing" workshop where service representatives from automobile manufacturers (domestic and import), aftermarket parts and equipment manufacturers, working technicians, and vocational educators meet in a workshop setting to share ideas and translate them into test questions. Each test question written by these experts must survive review by all members of the group.

The questions are written to deal with practical application of soft skills and system knowledge experienced by technicians in their day-to-day work.

All questions are pre-tested and quality-checked on a national sample of technicians. Those questions that meet ASE standards of quality and accuracy are included in the scored sections of the tests; the "rejects" are sent back to the drawing board or discarded altogether.

Each certification test is made up of between forty and eighty multiple-choice questions.

Note: Each test could contain additional questions that are included for statistical research purposes only. Your answers to these questions will not affect your score, but since you do not know which ones they are, you should answer all questions on the test. The five-year Recertification Test will cover the same content areas as those listed above. However, the number of questions in each content area of the Recertification Test will be reduced by about one-half.

Objective Tests

A test is called an objective test if the same standards and conditions apply to everyone taking the test and there is only one correct answer to each question.

Objective tests primarily measure your ability to recall information. A well-designed objective test can also test your ability to understand, analyze, interpret, and apply your knowledge. Objective tests include true-false, multiple choice, fill in the blank, and matching questions. ASE's tests consist exclusively of four-part multiple-choice objective questions.

The following are some strategies that may be applied to your tests.

Before beginning to take an objective test, quickly look over the test to determine the number of questions, but do not try to read through all of the questions. In an ASE test, there are usually between forty and eighty questions, depending on the subject. Read through each question before marking your answer. Answer the questions in the order they appear on the test. Leave the questions blank that you are not sure of and move on to the next question. You can return to those unanswered questions after you have finished the others. They may be easier to answer at a later time after your mind has had additional time to consider them on a subconscious level. In addition, you might find information in other questions that will help you recall the answers to some of them.

Do not be obsessed by the apparent pattern of responses. For example, do not be influenced by a pattern like **D, C, B, A, D, C, B, A** on an ASE test.

There is also a lot of folk wisdom about taking objective tests. For example, there are those who would advise you to avoid response options that use certain words such as *all, none, always, never, must,* and *only,* to name a few. This, they claim, is because nothing in life is exclusive. They would advise you to choose response options that use words that allow for some exception, such as *sometimes, frequently, rarely, often, usually, seldom,* and *normally.* They would also advise you to avoid the first and last option (A and D) because test writers, they feel, are more comfortable if they put the correct answer in the middle (B and C) of the choices. Another recommendation often offered is to select the option that is either shorter or longer than the other three choices because it is more likely to be correct. Some would advise you to never change an answer since your first intuition is usually correct.

Although there may be a grain of truth in this folk wisdom, ASE test writers try to avoid them and so should you. There are just as many **A** answers as there are **B** answers, just as many **D** answers as **C** answers. As a matter of fact, ASE tries to balance the answers at about 25 percent per choice **A, B, C,** and **D.** There is no intention to use "tricky" words, such as outlined above. Put no credence in the opposing words "sometimes" and "never," for example.

Multiple-choice tests are sometimes challenging because there are often several choices that may seem possible, and it may be difficult to decide on the correct choice. The best strategy, in this case, is to first determine the correct answer before looking at the options. If you see the answer you decided on, you should still examine the options to make sure that none seem more correct than yours. If you do not know or are not sure of the answer, read each option very carefully and try to eliminate those options that you know to be wrong. That way, you can often arrive at the correct choice through a process of elimination.

If you have gone through all of the test and you still do not know the answer to some of the questions, then guess. Yes, guess. You then have at least a 25 percent chance of being correct. If you leave the question blank, you have no chance. Your score is based on the number of questions answered correctly.

Preparing for the Exam

The main reason we have included so many sample and practice questions in this guide is, simply, to help you learn what you know and what you don't know. We recommend that you work your way through each question in this book. Before doing this, carefully look through Section 3; it contains a description and explanation of the question types you'll find on an ASE exam.

Once you understand what the questions will look like, move to the sample test. Answer one of the sample questions (Section 5) then read the explanation (Section 7) to the answer for that question. If you don't feel you understand the reasoning for the correct answer, go back and read the overview (Section 4) for the task that is related to that question. If you still don't feel you have a solid understanding of the material, identify a good source of information on the topic, such as a textbook, and do some more studying.

After you have completed all of the sample test items and reviewed your answers, move to the additional questions (Section 6). This time answer the questions as if you were taking an actual test. Do not use any reference or allow any interruptions in order to get a feel for how you will do on an actual test. Once you have answered all of the questions, grade your results using the answer key in Section 7. For every question that you gave a wrong answer to, study the explanations to the answers and/or the overview of the related task areas. Try to determine the root cause for your missing the question. The easiest thing to correct is learning the correct technical content. The hardest thing to correct is behaviors that lead you to a wrong conclusion. If you knew the information but still got it wrong there is a behavior problem that will need to be corrected. An example would be reading too quickly and skipping over words that affect your reasoning. If you can identify what you did that caused you to answer the question incorrectly you can eliminate that cause and improve your score. Here are some basic guidelines to follow while preparing for the exam:

- Focus your studies on those areas you are weak in.
- Be honest with yourself while determining if you understand something.
- Study often but in short periods of time.
- Remove yourself from all distractions while studying.
- Keep in mind the goal of studying is not just to pass the exam, the real goal is to learn!
- Prepare physically by getting a good night's rest before the test and eat meals that provide energy but do not cause discomfort.
- Arrive early to the test site to avoid long waits as test candidates check in and to allow all of the time available for your tests.

During the Test

On paper and pencil tests you will be placing your answers on a sheet where you will be required to color in your answer choice. Stray marks or incomplete erasures may be picked up as an answer by the electronic reader, so be sure only your answers end up on the sheet. One of the biggest problems an adult faces in test taking, it seems, is placing the answer in the correct spot on the answer sheet. Make certain that you mark your answer for, say, question 21, in the space on the answer sheet designated for the answer for question 21. A correct response in the wrong line will probably result in two questions being marked wrong, one with two answers (which could include a correct answer but will be scored wrong) and the other with no answer. Remember, the answer sheet on the written test is machine scored and can only "read" what you have colored in.

If you finish answering all of the questions on a test and have remaining time, go back and review the answers to those questions that you were not sure of. You can often catch careless errors by using the remaining time to review your answers. Carefully check your answer sheet for blank answer blocks or missing information.

At practically every test, some technicians will invariably finish ahead of time and turn their papers in long before the final call. Some technicians may be doing recertification tests and others may be taking fewer tests than you. Do not let them distract or intimidate you.

It is not wise to use less than the total amount of time that you are allotted for a test. If there are any doubts, take the time for review. Any product can usually be made better with some additional effort. A test is no exception. It is not necessary to turn in your test paper until you are told to do so.

Testing Time Length

An ASE written test session is four hours. You may attempt from one to a maximum of four tests in one session. It is recommended, however, that no more than a total of 225 questions be attempted at any test session. This will allow for just over one minute for each question.

Visitors are not permitted at any time. If you wish to leave the test room, for any reason, you must first ask permission. If you finish your test early and wish to leave, you are permitted to do so only during specified dismissal periods.

You should monitor your progress and set an arbitrary limit to how much time you will need for each question. This should be based on the number of questions you are attempting. It is suggested that you wear a watch because some facilities may not have a clock visible to all areas of the room.

Computer-Based Tests are allotted a testing time according to the number of questions ranging from one half hour to one and one half hours. Advanced level tests are allowed two hours. This time is by appointment and you should be sure to be on time to insure that you have all of the time allocated. If you arrive late for a CBT test appointment you will only have the amount of time remaining on your appointment.

Your Test Results!

You can gain a better perspective about tests if you know and understand how they are scored. ASE's tests are scored by American College Testing (ACT), a nonpartial, unbiased organization having no vested interest in ASE or in the automotive industry.

Each question carries the same weight as any other question. For example, if there are fifty questions, each is worth 2 percent of the total score. The passing grade is 70 percent. That means you must correctly answer thirty-five of the fifty questions to pass the test.

The test results can tell you:

- where your knowledge equals or exceeds that needed for competent performance, or
- where you might need more preparation.

Your ASE test score report is divided into content areas and will show the number of questions in each content area and how many of your answers were correct. These numbers provide information about your performance in each area of the test. However, because there may be a different number of questions in each content area of the test, a high percentage of correct answers in an area with few questions may not offset a low percentage in an area with many questions.

It should be noted that one does not "fail" an ASE test. The technician who does not pass is simply told "More Preparation Needed." Though large differences in percentages may indicate problem areas, it is important to consider how many questions were asked in each area. Since each test evaluates all phases of the work involved in a service specialty, you should be prepared in each area. A low score in one area could keep you from passing an entire test.

There is no such thing as average. You cannot determine your overall test score by adding the percentages given for each task area and dividing by the number of areas. It doesn't work that way because there generally are not the same number of questions in each task area. A task area with twenty questions, for example, counts more toward your total score than a task area with ten questions.

Your test report should give you a good picture of your results and a better understanding of your strengths and weaknesses for each task area.

If you fail to pass the test, you may take it again at any time it is scheduled to be administered. You are the only one who will receive your test score. Test scores will not be given over the telephone by ASE nor will they be released to anyone without your written permission.

3 Types of Questions on an ASE Exam

ASE certification tests are often thought of as being tricky. They may seem to be tricky if you do not completely understand what is being asked. The following examples will help you recognize certain types of ASE questions and avoid common errors.

Paper-and-pencil tests and computer-based test questions are identical in content and difficulty. Most initial certification tests are made up of forty to eighty multiple-choice questions. Multiple-choice questions are an efficient way to test knowledge. To answer them correctly, you must think about each choice as a possibility, and then choose the one that best answers the question. To do this, read each word of the question carefully. Do not assume you know what the question is about until you have finished reading it.

About 10 percent of the questions on an actual ASE exam will use an illustration. These drawings contain the information needed to correctly answer the question. The illustration must be studied carefully before attempting to answer the question. Often, technicians look at the possible answers then try to match up the answers with the drawing. Always do the opposite; match the drawing to the answers. When the illustration is showing an electrical schematic or another system in detail, look over the system and try to figure out how the system works before you look at the question and the possible answers.

Multiple-Choice Questions

The most common type of question used on ASE Tests is the multiple-choice question. This type of question contains three "distracters" (wrong answers) and one "key" (correct answer). When the questions are written effort is made to make the distracters plausible to draw an inexperienced technician to one of them. This type of question gives a clear indication of the technician's knowledge. Using multiple criteria including cross-sections by age, race, and other background information, ASE is able to guarantee that a question does not bias for or against any particular group. A question that shows bias toward any particular group is discarded. If you encounter a question that you are unsure of, reverse engineer it by eliminating the items that it cannot be. For example:

A rocker panel is a structural member of which vehicle construction type?

A. Front-wheel drive
B. Pickup truck
C. Unibody
D. Full-frame

Analysis:

This question asks for a specific answer. By carefully reading the question, you will find that it asks for a construction type that uses the rocker panel as a structural part of the vehicle.

Answer A is wrong. Front-wheel drive is not a vehicle construction type.
Answer B is wrong. A pickup truck is not a type of vehicle construction.
Answer C is correct. Unibody design creates structural integrity

by welding parts together, such as the rocker panels, but does not require exterior cosmetic panels installed for full strength.
Answer D is wrong. Full-frame describes a body-over-frame construction type that relies on the frame assembly for structural integrity.

Therefore, the correct answer is C. If the question was read quickly and the words "construction type" were passed over, answer A may have been selected.

EXCEPT Questions

Another type of question used on ASE tests has answers that are all correct except one. The correct answer for this type of question is the answer that is wrong. The word "**EXCEPT**" will always be in capital letters. You must identify which of the choices is the wrong answer. If you read quickly through the question, you may overlook what the question is asking and answer the question with the first correct statement. This will make your answer wrong. An example of this type of question and the analysis is as follows:

All of the following are tools for the analysis of structural damage **EXCEPT:**

A. height gauge
B. tape measure.
C. dial indicator.
D. tram gauge.

Analysis:

The question really requires you to identify the tool that is not used for analyzing structural damage. All tools given in the choices are used for analyzing structural damage except one. This question presents two basic problems for the test-taker who reads through the question too quickly. It may be possible to read over the word "**EXCEPT**" in the question or not think about which type of damage analysis would use answer C. In either case, the correct answer may not be selected. To correctly answer this question, you should know what tools are used for the analysis of structural damage. If you cannot immediately recognize the incorrect tool, you should be able to identify it by analyzing the other choices.

Answer A is wrong. A height gauge may be used to analyze structural damage.
Answer B is wrong. A tape measure may be used to analyze structural damage.
Answer C is correct. A dial indicator may be used as a damage analysis tool for moving parts, such as wheels, wheel hubs, and axle shafts, but would not be used to measure structural damage.
Answer D is wrong. A tram gauge is used to measure structural damage.

Technician A, Technician B Questions

The type of question that is most popularly associated with an ASE test is the "Technician A says . . . Technician B says . . . Who is right?" type. In this type of question, you must identify the correct statement or statements. To answer this type of question correctly, you must carefully read each technician's statement and judge it on its own merit to determine if the statement is true.

Sometimes this type of question begins with a statement about some analysis or repair procedure. This is often referred to as the stem of the question and provides the setup or background information required to understand the conditions the question is based on. This is followed by two statements about the cause of the concern, proper inspection, identification, or repair choices. You are asked whether the first statement, the second statement, both statements, or neither statement is correct. Analyzing this type of question is a little easier than the other types because there are only two ideas to consider although there are still four choices for an answer.

Technician A, Technician B questions are really double true or false questions. The best way to analyze this kind of question is to consider each technician's statement separately. Ask yourself, is A true or false? Is B true or false? Then select your answer from the four choices. An important point to remember is that an ASE Technician A, Technician B question will never have Technician A and B directly disagreeing with each other. That is why you must evaluate each statement independently.

An example of this type of question and the analysis of it follows.

A vehicle comes into the shop with a gas gauge that will not register above one half full. When the sending unit circuit is disconnected the gauge reads empty and when it is connected to ground the gauge goes to full. Technician A says that the sending unit is shorted to ground. Technician B says the gauge circuit is working and the sending unit is likely the problem. Who is right?

A. A only
B. B only
C. Both A and B
D. Neither A nor B

Analysis:

Reading of the stem of the question sets the conditions of the customer concern and establishes what information is gained from testing. General knowledge of gauge circuits and test procedures are needed to correctly evaluate the technician's conclusions. Note: Avoid being distracted by experience with unusual or problem vehicles that you may have worked on, Other technicians taking the same test do not have that knowledge, so it should not be used as the basis of your answers.

Technician A is wrong because a shorted to ground sending unit would produce a gauge reading equivalent to the test conditions of a grounding the circuit and produce a full reading.
Technician B is correct because the gauge spans when going from an open circuit to a completely
grounded circuit. This would tend to indicate that the problem had to be in the sending unit.
Answer C is not correct. Both technicians are identifying the problem as a sending unit but technician A qualified the problem as a specific type of failure (grounded) that would not have caused the symptoms of the vehicle.
Answer D is not correct because technician B's diagnosis is a possible cause of the conditions identified.

Most-Likely Questions

Most-Likely questions are somewhat difficult because only one choice is correct while the other three choices are nearly correct. An example of a Most-Likely-cause question is as follows:

The Most-Likely cause of reduced turbocharger boost pressure may be a:

A. wastegate valve stuck closed.
B. wastegate valve stuck open.
C. leaking wastegate diaphragm.
D. disconnected wastegate linkage.

Analysis:

Answer A is wrong. A wastegate valve stuck closed increases turbocharger boost pressure.
Answer B is correct. A wastegate valve stuck open decreases turbocharger boost pressure.
Answer C is wrong. A leaking wastegate valve diaphragm increases turbocharger boost pressure.

Answer D is wrong. A disconnected wastegate valve linkage will increase turbocharger boost pressure.

LEAST-Likely Questions

Notice that in Most-Likely questions there is no capitalization. This is not so with LEAST-Likely type questions. For this type of question, look for the choice that would be the LEAST-Likely cause of the described situation. Read the entire question carefully before choosing your answer. An example is as follows:

What is the LEAST-Likely cause of a bent pushrod?

A. Excessive engine speed
B. A sticking valve
C. Excessive valve guide clearance
D. A worn rocker arm stud

Analysis:

Answer A is wrong. Excessive engine speed may cause a bent pushrod.
Answer B is wrong. A sticking valve may cause a bent pushrod.
Answer C is correct. Excessive valve clearance will not generally cause a bent pushrod.
Answer D is wrong. A worn rocker arm stud may cause a bent pushrod.

You should avoid relating questions to those unusual situations that you may have encountered and answer based on the technical and mechanical possibilities.

Summary

There are no four-part multiple-choice ASE questions having "none of the above" or "all of the above" choices. ASE does not use other types of questions, such as fill-in-the-blank, completion, true-false, word-matching, or essay. ASE does not require you to draw diagrams or sketches. If a formula or chart is required to answer a question, it is provided for you. There are no ASE questions that require you to use a pocket calculator.

4 Overview of the Task List

Service Consultant (Test C1)

The following section includes the task areas and task lists for this test and a written overview of the topics covered in the test. The task list describes the knowledge and skills you should possess as an experienced service consultant. This is your key to the test and you should review this section carefully. We have based our sample test and additional questions upon these tasks. The overview section will also support your understanding of the task list.

ASE advises that the questions on the test may not equal the number of tasks listed; the task lists tell you what ASE expects you to know. The C1 test is unique in that it requires you to possess an understanding of customer service or "soft skills" as well as mechanical or product knowledge. The product knowledge sections of this guide will prepare you for the depth of knowledge determined, by a panel of industry experts, to be the minimum level required to adequately perform the job of service consulting.

At the end of each question in the Sample Test and Additional Test Questions sections, a letter and number combination will be used to direct you to the related task from which the question is derived for further study. Note the following example: **A.1.10.**

A. Communications (22 Questions)

Task A.1 **Customer Relations (16 Questions)**

Task A.1.10 **Identify and recommend service and maintenance needs.**

Example:

1. A vehicle in the shop for an oil change shows approximately 59,000 miles on the odometer. What should the service consultant do?
   A. Suggest an appointment for 60,000-mile maintenance.
   B. Offer the customer a discount to perform a 60,000-mile maintenance today.
   C. Advise that the 60,000-mile maintenance is covered under manufacturer's warranty.
   D. Provide a ball park estimate for a 60,000-mile maintenance. (A.1.10)

Analysis:

Question #1

Answer A is correct. Getting the customer prepared and asking for the appointment is the best. The customer may ask you to do the work while it is there. A 60,000-mile service is more than, but includes, an oil change. This is a major maintenance interval.

Answer B is wrong. Depending on the situation, you may be offering a discount to someone who would have the work done at regular price.

Answer C is wrong. This is not a warranty-covered item (with very rare exceptions).

Answer D is wrong. Guessing on prices can be a disaster. Customers always remember the lowest price in any range you give them.

Task List and Overview

A. Communications (22 Questions)

Task A.1 Customer Relations (17 Questions)

Task A.1.1 Demonstrate proper telephone skills.

Using the telephone properly includes greeting customers, speaking clearly, and demonstrating courtesy when taking calls or messages. Because of the varying environments in which service consultants work, questions addressing this "soft skill" are very "common-sense" based. Since the first word of the job title is service, use it as your guide on all areas of this test, excluding the product knowledge section.

Task A.1.2 Obtain and document pertinent vehicle information and confirm accuracy.

One of the most important communication skills the service consultant must perform is collecting the information needed to create a complete work order. Correct vehicle information is paramount throughout the repair process. Many updates and recalls are VIN-specific. Getting the correct parts is vehicle-specific. Even making sure that the right vehicle gets the right repair is driven by the repair order. Vehicle information includes year, make, model, powertrain info, VIN, production date, body configuration, and emissions or options information. Even issues such as whether it is a two-door or four-door, if it has power windows, or its color, can become essential.

Task A.1.3 Identify and document customer concern/request.

Interviewing the customer to determine their "concerns" is a skill critical in effective service writing. Properly documenting the conditions and descriptions of each concern should give the technician a good starting point in making a diagnosis. Simply writing down what the customer says is not enough. The skilled service consultant must be able to help the customer remember seemingly unrelated facts that may make the difference in a successful or unsuccessful diagnosis. This is done by asking questions that encourage the customer to verbalize the symptoms of their concern. Asking questions that start with when, how often, where, and to whom, can really help to fill in the question of why. As professionals, we must first fix the root cause for which the customer made the visit. This task also encompasses the collection of requested general service needs, like oil change or tire rotation. Questions throughout the customer relations area frequently overlap tasks in other sections. This is not a mistake; it is ASE's effort to test in different, real-world scenarios.

Task A.1.4 Obtain and document customer contact information.

Customer contact information may include name(s), phones, email, addresses, and faxes. Any way that the shop might contact the customer is fair game. Failure to contact at critical intervals of the repair process can lead to service mistakes and threaten shop credibility.

Task A.1.5 Open repair order and confirm accuracy.

The intention of this task is to use available repair order systems, either manual or computerized, to perform the administrative functions involved in starting the repair order process and checking to see that it is complete and correct. Due to the multitude of systems available, questions in this area will be very general in nature. This task will probably be grouped with another task to generate a real-world example.

Task A.1.6 Demonstrate appropriate greeting skills.

Greeting skills are the things we do and say to demonstrate to our customers that they are welcome. Shaking hands, smiling, verbal and non-verbal communications are all examples of greeting skills. ASE tests are built on industry standard procedures. Smiling and offering a handshake are common courtesy, common sense and industry standard.

Another similar example of a greeting skill would include asking the customer for their name. People go for service from people they like. Money can become the last issue. Eye contact is most important.

Task A.1.7 Arrange for alternative transportation.

One of the things service consultants are called on to do daily, is resolve transportation for customers. Options may include car rental or, at a minimum, a ride to home or work. The focus of this task is to test your understanding of the customer service and business liability side of the equation. Your shop might not have loaner vehicles, but it should be apparent to you that allowing an uninsured or unlicensed customer to drive a shop-owned vehicle would be a problem. Here, as in most of the soft skills areas, your ability to apply common sense is the most valuable tool you have.

Task A.1.8 Promote procedures, benefits, and capabilities of service facility.

A key to customer retention is the customer's understanding and perception of *value* in the work performed for them. A good service consultant sets his facility apart from the crowd by explaining the benefits of services and the skill level of the shop's technicians. Let us analyze the difference between features and benefits as it applies to this test.

An example of a feature might be longer-lasting spark plugs. The benefit is that the customer will not have to replace them as often. Most savvy customers buy benefits. An example of a feature with no real benefit is an extended warranty sold on a new car for 60,000 or 75,000 miles. This sounds like an added benefit, but the truth is that much is covered anyway by the standard warranty. Very little is extended. This is an example of selling on features. It might have sold some cars, but customers look for a feature to provide them with a convenience or some perceived value. Think of a benefit as something that provides a value in time, money, or convenience.

Task A.1.9 Review vehicle service history.

Vehicle history can mean the difference between a free repair and one a customer pays for. It can be an invaluable tool to help technicians in the diagnostic process. Questions in this task will be very general due to the myriad of ways that service facilities go about gathering history. If you understand the implications of ignoring history, you will be able to answer questions in this task. Computerized systems do a great job of tracking histories.

Task A.1.10 Identify and recommend service and maintenance needs.

There are many possible ways to identify service and maintenance needs. As a service consultant, you may not be using or have access to all of them, but should be aware of them. The most basic method is looking at the vehicle's odometer. Customer history can play a role here. The vehicle's owner manual contains much of the maintenance information. Manufacturer and aftermarket information systems and web sites, as well as the National Highway Traffic Safety Association, are some of the electronic means available. The last and probably most viable source is the technician who conveys maintenance needs on the repair order.

Task A.1.11 Communicate completion expectations.

In a recent AAA survey, the number one complaint customers had with repair shops was that their vehicles were not completed at the time they were promised. Anticipate that questions will address insuring and communicating completion expectations. For many service consultants, promising a completion time at the time of drop off is standard operating procedure. The industry is sending the message that the best "communication" occurs after the technician evaluates the vehicle. He must update the customer throughout the repair if timely, and avoid over promising.

Task A.1.12 Obtain repair authorization.

Taking into consideration estimate laws in various states, this area of the test revolves around providing the customer with pricing, asking for the sale, and documenting authorization for needed repairs. To enter or drive the car in the shop is a risk without a signature.

Task A.1.13 Identify customer type (first time, warranty, repeat repair, fleet, etc.).

The task pretty well describes this. The questions will revolve around the "why" question. Why would you want to identify the customer type? As we all know, customers often use the word "they" to describe anyone who worked on their vehicle. Does "they" mean another shop, many other shops, or is it an indirect way of referring to those invisible "theys" that work in your shop? If you have any question about why we identify the customer type, re-read the description of Task A.1.9 or try to figure out which one of the 49 identical trucks to choose from your computer system when an unidentified fleet customer visits for an oil change.

Task A.1.14 Present professional image.

Presenting a professional image can be as varied as how we dress to the way we treat or talk to customers and other employees. As a service consultant, you are often the only example the customer has to generate an opinion of the quality of your facility.

Because you are the point person for your business, the likelihood of a customer returning to your facility by choice lies more on your shoulders than anyone else's. The grooming of your facility is also a key component of presenting a professional image to your customer. Cleanliness of yourself and your facility plays a big part in your image evaluation.

Task A.1.15 Perform customer follow-up.

Pay attention to customer follow-up; it is a critical area of customer service communication. It can go from making a call to a customer to thank them for bringing in their vehicle, to taking the responsibility of calling when a special order part arrives. Remember birthdays with a card or send thank-you cards on Thanksgiving. Add value and create trust through caring.

Task A.1.16 Explain and confirm understanding of work performed, and charges; review methods of payment.

The shop is crazy and five people are waiting for you. The temptation is to throw the keys and repair order on the desk and mutter something about the vehicle being "fixed" while you cash-out the customer. The service consultant has the responsibility of controlling all customer contact so that a complete understanding is reached. The time saved with a quick pick-up can be totally offset by the angry follow-up that occurs when the customer's perceptions do not match the actual work performed. The message here is that five minutes spent can save you hours, or even save you a *customer*. Plus, all of those customers waiting get to see what a really nice person you are before you are face-to-face with them. The pick-up is the last and sometimes the only shot you get for a face-to-face marketing effort. Remember that people do not like having their vehicle repaired, but that doesn't mean that they cannot like you. When possible, try to schedule pick-up times. This can avoid the stress that develops when all of the customers show up at once.

Task A.2 Internal Relations (6 Questions)

Task A.2.1 Effectively communicate customer service concern/request.

Collecting all the information in the world from Task A.1.3 is of no value at all unless the service consultant can make sure that the information collected is useful to the technician. Effectively communicating can also mean not having to call the customer back several times. Expect questions in this task to focus on the communication between the service consultant and the technician relating to work orders. In the future, it may be interpreted to include verbal communication. Do not add or delete any information. All may be pertinent.

Task A.2.2 Understand the technician's diagnosis and service recommendations.

Arguably, the real craft of service consulting is the ability to take the diagnosis and recommendations of a technician and turn it into a coherent description of the "cause" and "correction" needed to satisfy and alleviate the customer's "concern." This area is really about your product knowledge put to practical work. To that end, you really must have your "mechanical" knowledge together. For most service consultants, this means asking questions of technicians to gain an understanding of how the various systems work. The most successful service consultants do an excellent job of taking the technical and offering an accurate but customer-friendly explanation of the problem. Remember, most customers will not want an in-depth technical explanation. Keep it simple and to the issue.

Task A.2.3 Verify availability of required repair parts.

Whether your environment has a parts department or you are responsible for parts ordering, you will always find yourself on the firing line when parts problems occur. To avoid problems, the skilled service consultant confirms the availability and timing of parts before promising a return time to the customer or directing the technician to begin work. The magic operative is communication. Make notes for yourself and reminders if you are expecting a part at a certain time and hold your suppliers to the promises they make so that you may keep yours to your technicians and customers. This task has a sort of procedural aspect to it, so expect to have your common sense tested here. If the part supplier tells you that the fuel pump you need will not be on your doorstep until 3:00 P.M., Service Consultant A says you don't promise the vehicle for 2:30. Service Consultant B says if you do, you are going to have a technician, your customer, and possibly your service manager using your name in vain.

Don't expect to see the use of purchase orders or any specific parts system in here since not all businesses are the same. Do expect that you will need to make the mental leap to scenarios that might occur if you work with a parts department or get your parts from an outside supplier. Communication becomes key in this area.

Task A.2.4 Establish completion expectations.

If your customers are setting your schedule, this will come as a surprise to you. Service consultants working with technicians are more able to determine the completion time of any given task. Many consultants have a very good knowledge and working relationship with their technicians and can plan workflow with excellent results. Most technicians would prefer to be given the order in which vehicles need to be returned to their owners. All of the other factors, such as customer need, parts availability, and technician/equipment availability play into this. Some jobs require different skills which should also be considered job distribution. The consultant must take the lead working either with shop supervisors or technicians to offer customers realistic completion times.

Task A.2.5 Monitor repair progress.

During the course of the day, maintaining a hands-on approach to work flow keeps the shop's productivity high. You will have more options if problems that come up can be addressed and communicated to the customer as early as possible. You may find that a walk through the shop to gather updates from your technicians can be a simple and effective way to monitor repair progress. Some electronic dispatching systems allow members of the team to update their progress throughout the day. As consultant, you play the middle, between the technician and the customer.

Task A.2.6 Interpret and clarify repair procedures.

This area is going to heavily overlap other tasks. It provides for opportunity to test your ability to understand the purpose of a repair procedure. This task will probably be used along with customer communication tasks.

Task A.2.7 Document information about services performed or recommended.

When the shop performs a service, it is critical to document information in such a way that the technician could review notes and know what work was performed and why. It is also important to document in such a way that the customer understands and can perceive value in the service provided. This may include maintaining internal documents along with final customer invoices or providing references to TSB (technical service bulletins) relevant to the work performed. If a recommended, but not repaired, service is not documented, a customer may blame the shop for failure.

Task A.2.8 Communicate with shop personnel about shop production/efficiency.

The difference between productivity and efficiency, as well as how to keep things humming along, are the focus of this task. Productivity is the amount of work a shop gets out the door while efficiency is a measurement of how effectively work is completed. If a technician is at work for 8 hours, has 8 hours of sold labor, and completes it in 8 hours, he is 100% effective and 100% productive. If, on the other hand, he is at work for 8 hours, has 6 hours of sold labor, and completes the work in 4 hours, he is 150% effective but only 75% productive. This information provides an invaluable tool for tracking technicians. In our first scenario, the technician and the service consultant are working in harmony. The only way to make more in this scenario is for the technician to have higher efficiency so that more work can be completed in the same amount of time. This often calls for scheduling some highly efficient work that can be performed faster than the "book" time because of the technician's experience and skill. This relation between productivity and efficiency is critical to the profit of the shop and financial success of the employee.

In our second scenario, our tech is working, but does not have enough work. If you shuffle the numbers around, you can see that this tech might easily produce 12 hours a day if the work were there. Questions to this task will be pretty general as most service consultants in the OE Dealer networks are not called on to perform this kind of analysis. This is invaluable information to understand in any service environment because it is the difference between happy technicians, profitable shops, and the revolving employee/business failure scenarios.

Task A.2.9 Maintain open lines of communication within the organization.

The last task in our internal relations section may well be the most ambiguous, since maintaining open lines of communication could encompass every kind of communication in the business. It seems that most of the questions written to this task will involve communication between individuals. This might include verbal, written (memo), or telephone messages. Expect that problem resolution will be a popular subject. Small meetings daily, weekly, or monthly can help every employee have a chance to contribute to finding remedies to problems.

B. Product Knowledge (27 Questions)

Sections B.1 through B.7 cover mechanical knowledge of vehicle systems. Because B.1 through B.4 sections have the same three tasks, we will address all three tasks at the same time. The tasks are: 1) Identify major components and location, 2) Identify component function, and 3) Identify related systems. Each section is organized with charts and diagrams to help you get a working knowledge of each system, including how they relate to other systems. The first task asks that you identify the location of major components. It would be better if it were asking you to identify the relationships to other components. Vehicles have so many design variations it would be nearly impossible to write a question that would always fit every vehicle built. We will be using the word "usually" in the mechanical sections for this reason. ASE test questions do not use words like "always" because of this rule. The service consultant is not expected to possess the depth of knowledge that a technician must, but a working knowledge of system functions is necessary to provide good customer service and to communicate with technicians and parts people.

1. Tasks 1, 2, and 3 Engine Systems (7 Questions) (includes mechanical, cooling, fuel, ignition, exhaust, emissions control, and starting/charging).

An integrated approach to each system indicated on the task list should help you to gain a better understanding of the components and their relationships. For each number in the exploded diagram, there is a corresponding numbered section in the chart following it. If you understand this information you will have adequate knowledge to answer any questions on the Service Consultant Test.

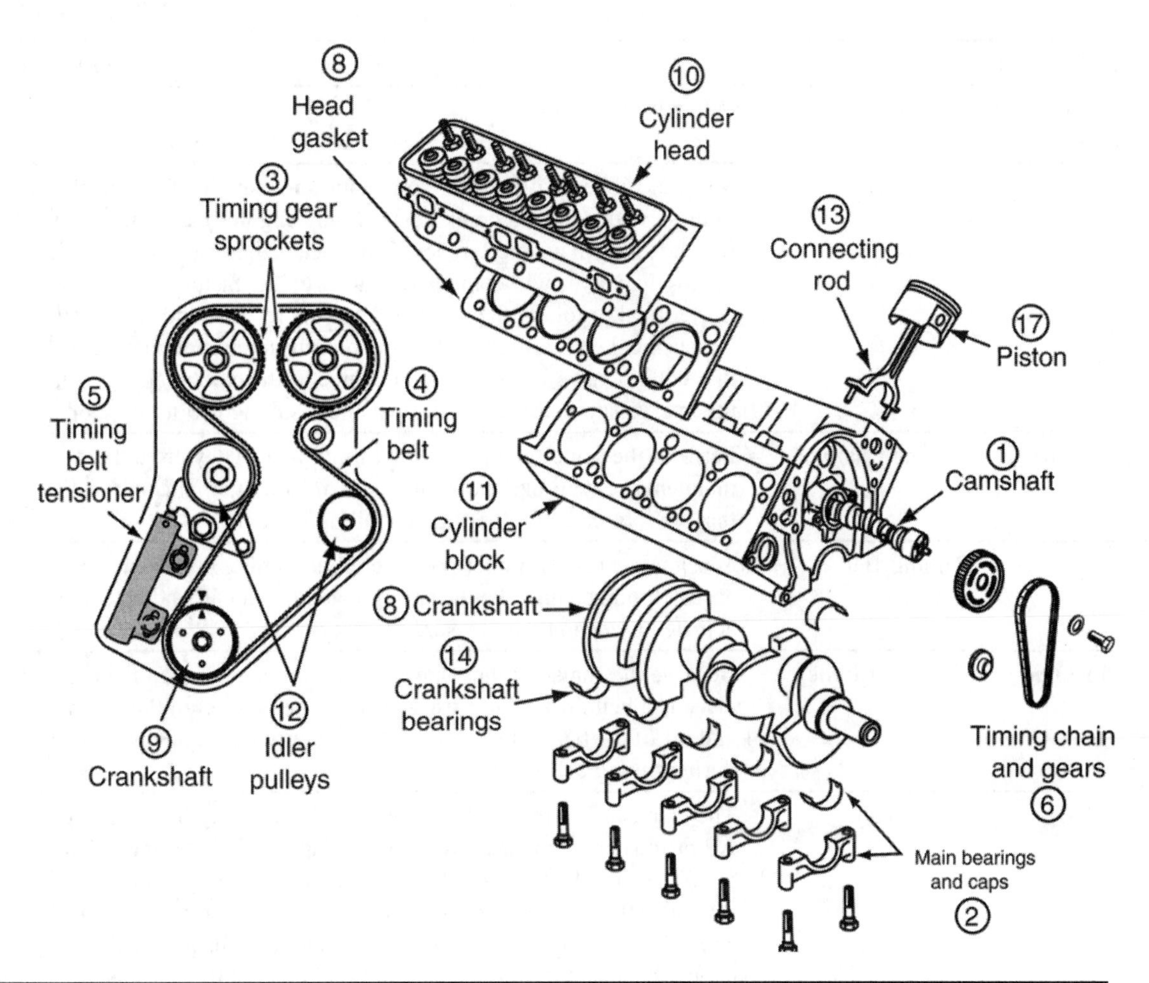

VALVE TRAIN		
Diagram Number	**Component Name**	**What It Does**
1.	Camshaft	The camshaft has eccentric lobes that rise off of the centerline of the shaft to push on valve lifters or rocker arms and cause valves to open. The cam is driven at half crank speed by timing chain, timing belt or timing gears connected to the crankshaft.
2.	Main Bearings	These attach the crankshaft to the block. Must be torqued to and Caps specification. A sequence of tightening may be required.
Not Pictured	Valves	Intake valves control the introduction of fuel and air into the engine and open at the beginning of the intake stroke. Exhaust valves control the elimination of by-products of the combustion process. Both valves are closed during the Compression and Power strokes.
3. or 6.	Timing Gears	These are driven by the timing belt or a chain. The example shown is a belt-drive design dual-overhead camshaft. This gear is many times referred to as a sprocket.

VALVE TRAIN		***(Continued)***
Diagram Number	**Component Name**	**What It Does**
4.	Timing Belt	Made of rubber, with reinforcing bands and teeth made of rubber, to mesh with cam and crank gears. Many engines have valvetrain, which is known as interference. This means that if the timing belt breaks the valves will come in contact with the pistons. This is why it is critical to follow the mileage intervals recommended for belt replacement. It is still important in non-interference engines (engines that will not have valve-to-piston contact if the belt breaks) to follow the intervals to avoid a break down and resulting tow to the shop.
Not Numbered	Timing Chain and Gears	Perform the same function as the timing belt. They are not a maintenance item like the timing belt and are internal components that are lubricated by the engine's oiling system.
5.	Timing Belt Tensioner	Tensioners provide proper pressure on the timing belt or, in some cases, timing chain. May be spring loaded, manually adjusted, or actuated by engine oil pressure.
Not Pictured	Camshaft Pulley Seals	Because the camshaft must have oil to lubricate it, when the cam drive is a belt drive, there are seals that keep the oil in the engine and off of the belt. This I s a very commonly replaced component during timing belt service.
7. and 9.	Crankshaft	The crankshaft is the rotating member that provides power output. All pistons are connected by a connecting rod and bearing. The crankshaft is a large offset shaft. The journals are configured in 60-, 90-, or 180-degree relationships depending on the type of engine. The large weights hanging opposite of the journals are used to counterbalance the weight of the journal and the rod and piston assembly that attaches to each journal. In most V-6 or V-8 engines, there are two connecting rods bolted Siamese to each crank journal, while 4-cylinder engines usually have one rod on each crank journal. This is a heavy component made of steel or nodular iron in most cases. This rotates at twice the speed of the camshaft.
8.	Head Gasket	Provides a positive seal between the head and cylinder block. Oil, antifreeze, and combustion are all sealed by this component part.
Not Pictured	Front Crank Seal	This component functions the same as a cam seal. You will find a front crank seal in all engines regardless of valvetrain configuration. This component often fails and is replaced during timing belt service.
Not Pictured	Valve Springs and Retainers	These apply pressure to keep the opening and closing of the valve in matching motion to the lobes of the camshaft and to hold them closed.
Not Pictured	Timing Cover	Used to protect the timing belt. In timing chain applications it covers the timing chain, controls oil, and often contains the oil pump.
Not Pictured	Valve Cover	The cover that attaches to the cylinder head and covers the valvetrain components.
10.	Cylinder Head	Holds all of the valvetrain components and has air flow ports for the exhaust and intake. The cylinder head bolts onto the cylinder block and has the intake and exhaust manifolds bolted to it. The complete package is often called the induction system.

SHORT BLOCK ASSEMBLY		
Diagram Number	**Component Name**	**What It Does**
11.	Cylinder Block	The backbone of the engine; all major components bolt to it. The cylinder bores are the large holes in it. The pistons go in these holes. The crankshaft rotates in bearing inserts in the block's main saddles.
12.	Idler Pulleys	These simply provide direction change in the belt configuration. They are in a fixed mounting and simply "idle" in rotation when the belt is in rotation.
13.	Connecting Rod	The connecting rod has two holes in it when viewed by itself. The large hole is the side that holds the rod bearing and splits apart to bolt to the crank journal. The small end accepts the wrist pin of the piston, which is the point the piston pivots on when the crankshaft is spinning.
14.	Crank Bearings	The bearings in modern engines are a composite of different materials clad together. The back shell is usually aluminum with a copper layer on it, and then a soft metal alloy, similar to lead, that is the actual bearing surface. The bearing does not have any moving parts like other bearings. In the engine its job is to carry an oil film to the surfaces of the crank and maintain adequate clearance between moving parts. When the clearances are excessive, knocking noise and loss of oil pressure can result. If the bearing clearance is too tight, the oil film that cools and protects the parts is too thin to be effective, which will cause damage of the parts.
Not Pictured	Oil Pump	The heart of the engine. Oil pumps may be driven directly by the crankshaft or by gears and a drive shaft from the valvetrain. The pump pulls oil from the oil pan, passes it to the oil filter. Because it can move more volume than the internal clearances of the engine can flow, pressure is created in the oiling system. If these clearances increase due to wear in the bearings due to age, oil flow may increase, but the pressure will be lowered. This pressure occurs in small passages inside the engine, sort of like the plumbing inside a house, carrying oil to all of the bearing surfaces in the engine. The oil pan and valve covers are not under pressure. They simply keep the oil in the engine. The oil pan is a reservoir for the oil to return to. Engines use drains like the gutters on a house to return unpressurized oil back to the pan at the bottom of the engine.
Not Pictured	Rear Main Seal	This is the largest seal that keeps the oil from the rear crank main bearing inside the engine. This seal may be a two-piece design that is serviced by removing the rear main bearing cap, or it may be a one-piece full circle seal that is installed from the outside rear of the engine, requiring transmission and/or transaxle removal.

SHORT BLOCK ASSEMBLY		***(Continued)***
Diagram Number	**Component Name**	**What It Does**
17.	Piston	The piston is made of aluminum in all modern engines. When it moves to the bottom of the cylinder it creates a vacuum that pulls air and fuel into the cylinder; this is called the intake stroke. It then rises to the top to compress the air it just took in, called the compression stroke, before the spark plug sparks and causes the high-pressure fuel and air mixture to create lots of heat as it oxidizes (burns). This heat causes expansion, which forces the piston back down. This is the power stroke. This causes the crank to spin because of those eccentric journals we discussed earlier. The crank continues to spin on inertia and we get our final stroke in the 4- stroke cycle know as the exhaust stroke. This is where we get rid of the leftover by-products of combustion. Since we have different cycles happening in the other cylinders, the crank is being pushed by a compression stroke from another piston continuously. Due to the 4 strokes needed, the crankshaft will rotate twice for every camshaft rotation. This is why the crank gear will always be the smaller, or half size, when compared to the camshaft gear.
Not Pictured	Piston Rings	There are two types of piston rings on each piston. One is the compression ring. There are usually two of these stacked right on top of each other in grooves in the piston. Since the piston must be lubricated to slide up and down the cylinder, it cannot be too tight in the bore. The compression rings are responsible for providing an extremely effective and small sealing surface that requires very little lubrication. The second ring type is the oil ring, which helps to carry oil up the cylinder bore to lubricate and to scrape it back off and pass it through notches cut in the piston skirt on the way back down.

Cooling System

The cooling system serves two purposes. First, it keeps the engine at a steady temperature. Second, it provides the hot water that makes the heater work. The cooling system is made up of many components. The diagram below will show a basic system common to most vehicles. The cooling system contains a mixture of antifreeze and water. The mixture carries heat from the engine components to the radiator and the heater core where the heat is exchanged by convection between the outside air and the radiator or heater core. Take a look at the components and how they work together.

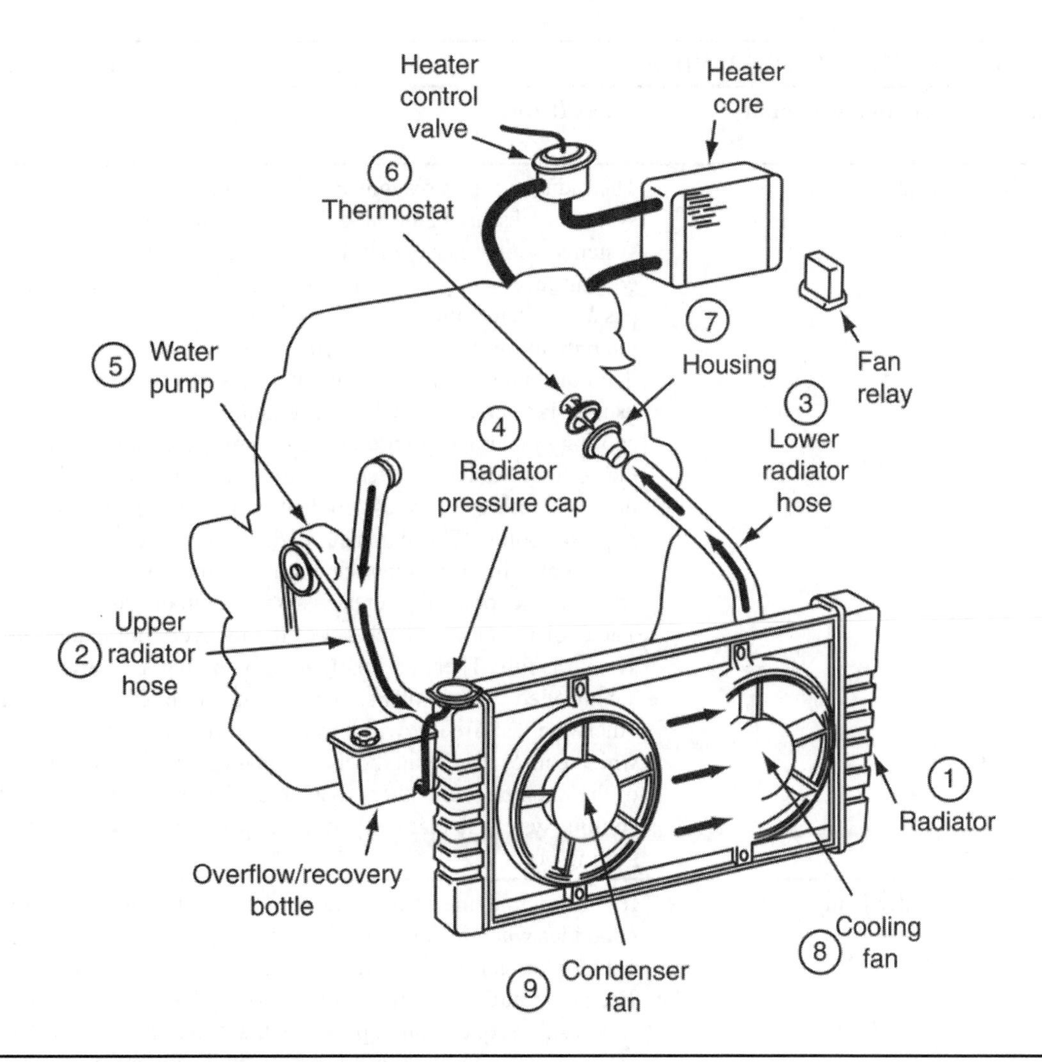

COOLING SYSTEM EXPLODED VIEW

Diagram Number	Component Name	What It Does
1.	Radiator	At its very simplest the radiator is a heat exchanger. Warm coolant enters it, the outside air moving across it removes heat from the coolant, and it returns to the engine to start the cycle over again. Radiators are carefully chosen to have the correct amount of heat transfer efficiency called BTU (British thermal units) If the radiator is too small or becomes dirty or restricted, the engine will overheat due to inadequate heat exchange. It should be noted that most vehicles equipped with automatic transmissions circulate their transmission fluid through a separate heat exchanger inside the radiator to help to warm the transmission up on cold days and to maintain consistent temperature after the vehicle has warmed up.
2.	Upper Radiator Hose	The radiator hoses are rubber hoses that provide a flexible connection between the engine and radiator.
3.	Lower Radiator Hose	The radiator hoses are rubber hoses that provide a flexible connection between the engine and radiator.

COOLING SYSTEM EXPLODED VIEW *(Continued)*

Diagram Number	Component Name	What It Does
4.	Radiator Cap	The radiator cap is responsible for maintaining the cooling system's pressure. Late model vehicles have closed cooling systems, which means that they use a radiator cap to control the coolant level along with a recovery bottle. The radiator cap has a seal that comes in contact with a surface inside the top of the radiator or coolant reservoir. There is a spring in the cap that holds pressure against this sealing surface. The pressure is usually between 13 and 16 psi. When the engine is cold the cooling system has no pressure in it. As the engine warms up, the coolant expands. When the pressure in the system exceeds the cap's rating the excess coolant will be pushed out to a recovery bottle. This expansion will also push any air in the system out. In turn, when the engine is shut down the coolant retracts as it cools. This creates a vacuum in the cooling system. The coolant is drawn back out of the recovery bottle until the system is full. There are really no moving parts involved; just a controlled use of expansion and contraction. If the system is otherwise leak free, it will be full at all times and free of air which causes accelerated deposits that can restrict the tubes of the radiator. One other advantage of running a pressurized cooling system is that coolant under pressure has a higher boiling point.
5.	Water Pump	If you can imagine an old paddle wheel boat, you have a pretty good idea what the inside of a water pump looks like. One of the radiator hoses is connected to the inlet side of the water pump. The pump pulls coolant from the cooled side of the radiator and pushes it out through passages that travel through internal passages in the cylinder head and block. These passages are cast around all of the hot parts of the combustion process, like the cylinders and the combustion chambers in the heads. The shaft that connects to the "paddle wheel," called an impeller, goes to the outside of the engine and has a pulley attached to it that is driven by a belt. This shaft has seals and bearings to allow it to spin at high speed while keeping the coolant in the engine. This is the location most water pumps fail. Water pumps come in every shape and size but all perform the same task.
6.	Thermostat	Probably the most misunderstood component of the cooling system, the thermostat does not keep the engine cool, it warms it up. The radiator is responsible for keeping the engine cool. To ensure that the engine wears evenly and produce the least amount of tailpipe emissions it must be kept at a constant and even temperature throughout. Engines make more excess heat when they are under load than they do when they are cruising under light loads. The radiator is a static device that is more efficient when there is more air flowing across it. Since ambient temperatures and the amount of air moving across the radiator can vary, something has to control it. This is the job of the thermostat. For most people, the thermostat is a device with a control knob that adjusts the temperature of their home. Here we have to get a different visual of the device. The thermostat

COOLING SYSTEM EXPLODED VIEW		*(Continued)*
Diagram Number	**Component Name**	**What It Does**
		for an engine is located inside the engine and submerged in coolant. It controls the amount of coolant that enters the radiator from the engine. When the engine is cold, a big spring inside the thermostat holds it closed to keep all of the coolant flowing in the engine. As the engine reaches operating temperature, usually about 195° F, the spring in the thermostat relaxes, allowing coolant to flow to the radiator for heat exchange. If the temperature drops, the thermostat closes the opening down. This keeps the temperature consistent.
7.	Thermostat Housing	Most engines have a thermostat housing which is either part of the engine or a separate part. They are often made of light duty aluminum and have a radiator hose connection. They are subject to leaks and warpage in many applications.
8.	Cooling Fan	Since vehicles operate under a variety of conditions there is no way to guarantee adequate airflow across the radiator, for instance when sitting in traffic or moving at slow speeds. The cooling fan is used to provide the wind when there is none. Cooling fans may be the old static fans that moved at the same speed as the water pump they were bolted to, they may be electric and controlled by a temperature switch or engine computer, or they may be mounted on a fan clutch.
9.	Fan Motor	In electric fan applications a motor is used to spin the fan.
Not Pictured	Fan Clutch	Fan clutches are used in applications where the fan is mounted to a moving part of the engine, usually the water pump. Fan clutches can be thermostatic or speed sensitive. In either case they vary the amount of air the fan moves by slowing down the fan blade relative to engine speed. This allows the use of a large, highly effective blade for slow speeds that can be slowed down to improve engine performance when it is not needed.
10.	Heater Core	The heater core is a small radiator. Coolant from the engine is circulated through it, and a blower blows across the core to provide hot air inside the vehicle.
11.	Overflow/Recovery Bottle	Coolant recovery/overflow bottles are the reservoirs that coolant moves into and out of as it expands and contracts. In many late model vehicles, the bottle has been replaced by a totally closed system where the bottle functions as the radiator pressure and off gassing tank to help trapped air get out of the coolant.

COOLING SYSTEM EXPLODED VIEW		*(Continued)*
Diagram Number	**Component Name**	**What It Does**
Not Pictured	Antifreeze	The coolant in the engine comes in many types. It can be green, red, yellow, orange, or pink. The name antifreeze is only half the story. In addition to dropping the freezing point of water to around −40° F, it also helps to prevent boiling clear up to 240° F or more, depending on system pressure. The coolant is mixed with water in a 50/50 mixture in almost all applications. The coolant has the additional responsibility of lubricating the water pump, protecting the metal components of the engine, and improving the temperature transfer capabilities of the components it comes in contact with.
12.	Fan Relay	Relay Provides a low-current control for a high-current demand (fan motors). On modern autos this cycle is computer controlled, receiving inputs from many sensors.

Fuel Systems

We are going to take the fuel system in two parts. First, we will look at the induction system. This system manages, measures, and directs airflow into the engine. Next, we will look at the fuel components of a fuel-injection system. We will not be providing information on carbureted systems, as they were not included in the fuel system section of the Service Consultant Test.

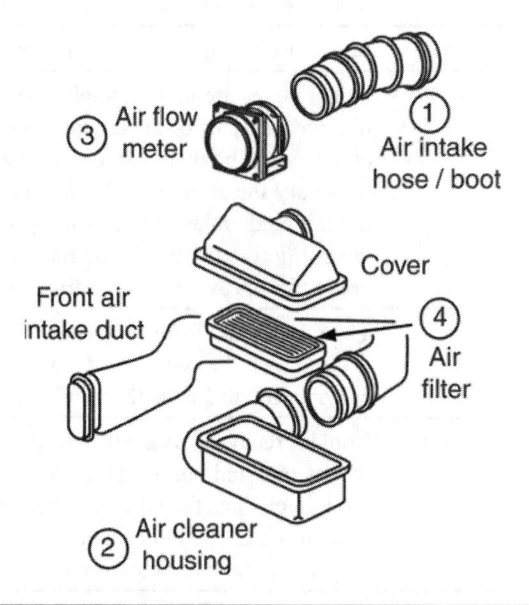

INDUCTION SYSTEM		
Diagram Number	**Component Name**	**What It Does**
1.	Air Intake Hose/Boot	This hose connects the air cleaner to the throttle body in most fuel injected vehicles. It may have connections to the PCV system. Because it provides the flex between the body and the engine, when breakage occurs it causes strange drivability problems due to false, non-calculated air entering the system.

INDUCTION SYSTEM		*(Continued)*
Diagram Number	**Component Name**	**What It Does**
2.	Air Cleaner Housing	Holds the air filter. There are as many different configurations as there are vehicles but they all hold the air filter. Some manufacturers mount the air meter or mass airflow sensor on the air cleaner as in this picture. Only filtered clean air can flow past the air flow meter.
3.	Air Flow Meter	There are two types of meters commonly in use and they include the vane type and the mass air flow meter. The vane type meter has a moving flap in the incoming air tract that opens and closes based on how much air the engine is using. The computer uses this signal to calculate how long to open each injector. The other type of meter uses a heated wire across the intake stream. When air crosses this wire or grid the wire cools off. This change in temperature is used by the computer to calculate how much air the engine is using and how much fuel to deliver.
4.	Air Filter	The air filter element has the job of removing dirt and other contaminants from the incoming air to the engine. All of the various designs perform this same function. These are replaced usually at mileage intervals. It's always best to know and understand your customer. A severe mileage interval may require more frequent change.

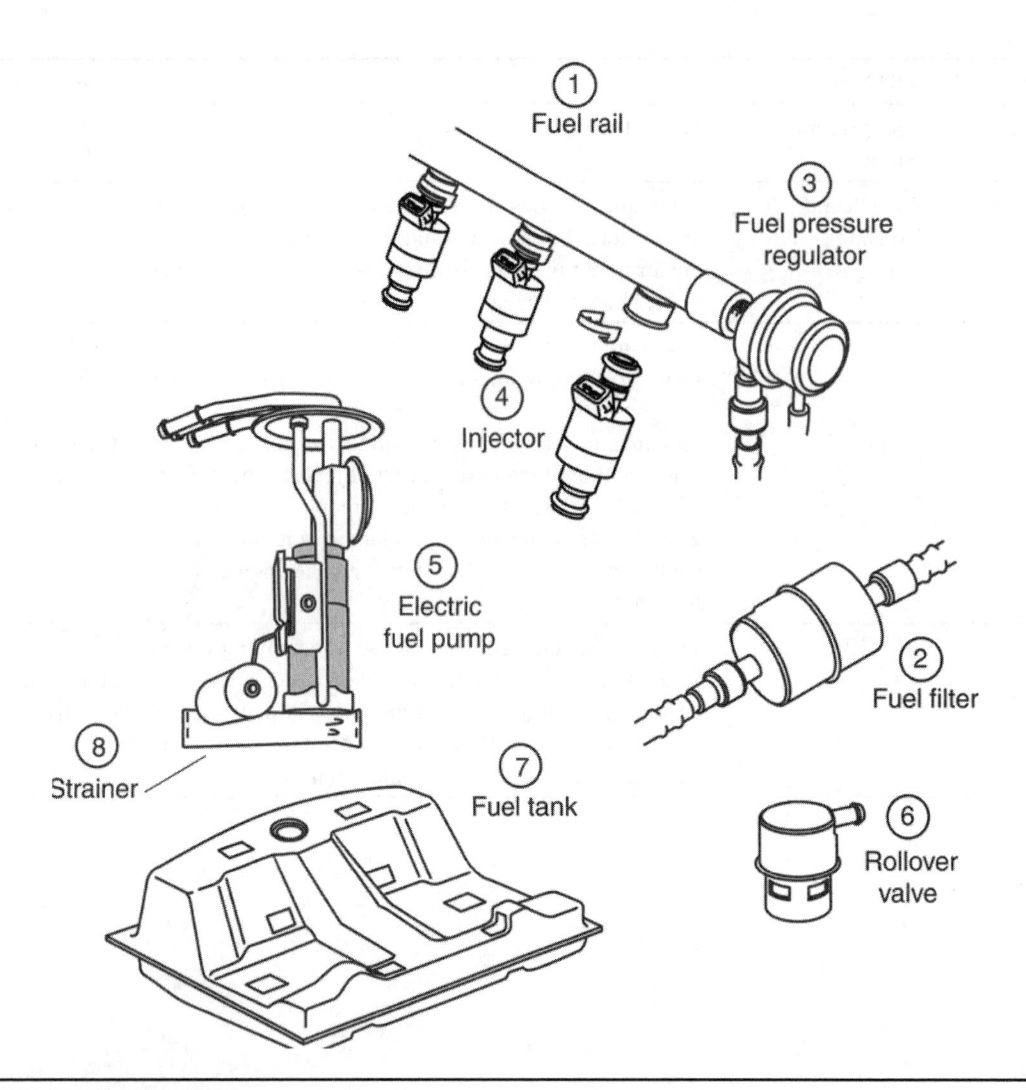

FUEL INJECTION SYSTEM

Diagram Number	Component Name	What It Does
1.	Fuel Rail	The fuel rail connects the fuel lines to the fuel injectors. In many applications the fuel-pressure regulator is attached to the fuel rail. The fuel injectors are retained in the fuel rail by O-rings and, in some applications, clips. This is traditionally not a part that wears out.
2.	Fuel Filter	Fuel filters contain the filtering media used to remove dirt and debris from the fuel system after the fuel pump but before the fuel injectors.
3.	Fuel-pressure Regulator	The fuel-pressure regulator is located in the engine compartment Regulator most of the time. It resides in the return side of the fuel line and keeps the fuel pressure at the necessary level by restricting the amount of fuel returned to the tank. The control for the pressure may be a fixed spring inside the regulator or vacuum from the engine. In a returnless system, fuel pressure may be controlled by the PCM. These systems do not have a return line but rather vary the speed of the electric fuel pump. In this situation the regulator is a monitoring device telling the computer how much pressure is being made.

FUEL INJECTION SYSTEM		***(Continued)***
Diagram Number	**Component Name**	**What It Does**
4.	Fuel Injector	A fuel injector is an electrical component. It is a high-speed solenoid. When the computer completes the electrical circuit to the pintle, which is a small needle valve, it is lifted up off the seat and fuel sprays out of the tip in a nice cone shape. When the computer releases the circuit the valve closes. This is measured by the technician in milliseconds. An injector is typically open for 1–2 milliseconds at idle and 12–15 milliseconds under heavy load. This rapid movement and the cycle from cold to hot and back again are the most common causes of injector failure.
5.	Fuel Pump Module	The fuel pump module houses the fuel pump and, in many cases, the fuel gauge sending unit along with any plumbing necessary to complete the fuel return to the tank. Most manufacturers sell this as an assembly. Some fuel pumps are sold separately and are installed in the original module or mounted in other places on the vehicle. The in-tank unit is, by far, the most common design.
6.	Rollover Valve	It is likely that you have not run into this component. This device serves two purposes. The first is to keep fuel from running out of the fuel tank vent line in the event of a vehicle roll over. The second is a check valve for the evaporative-emission system. Evaporative emissions systems in fuel-injected vehicles are really designed to control fuel vapors that might leave the tank as the engine warms the fuel from the constant circulation through the fuel rail and back to the tank. In vehicles built from 1996 on, the evaporative system is controlled and monitored by the engine-management computer or Powertrain control module (PCM). The fuel tank is a critical and major component of this system. The largest evaporative system line is usually connected to the roll-over valve and then to the engine on the other end. A solenoid-controlled vacuum valve is used by the PCM to apply vacuum to the tank for testing and to draw vapor into the engine to be burned.
7.	Fuel Tank	In late model vehicles, the fuel tank is more than just a holding compartment for fuel. It functions as a surge tank to manage the changes in fuel temperature. As we mentioned before, it usually contains the fuel pump and a sending unit. It also has baffles inside to help the fuel stay as close to the fuel pump pick up as possible. 8 Strainer Prevents large debris from being drawn into the fuel pump and likely being pushed forward to the fuel filter.

Ignition System

The ignition system is an electrical system. It is made up of two parts. The one most obvious to our customers is the part that includes the ignition switch. This part of the ignition system has grown from a simple switch to turn the power on to the vehicle engine to the master control switch for the entire vehicle. For our purposes, you will have enough knowledge to know what the ignition switch does. The second part of the ignition system is what technicians have traditionally called the ignition system. This includes the components that have been the tune-up parts. Ignition components provide the high-voltage, low-amperage spark to fire the combustible gases to start. With vehicle maintenance seeing so many changes in the last decade, the tune-up has become a very gray area that does not really fit anymore. In the next area, we are going to look at the components of both a conventional distributor electronic ignition and distributorless ignition system.

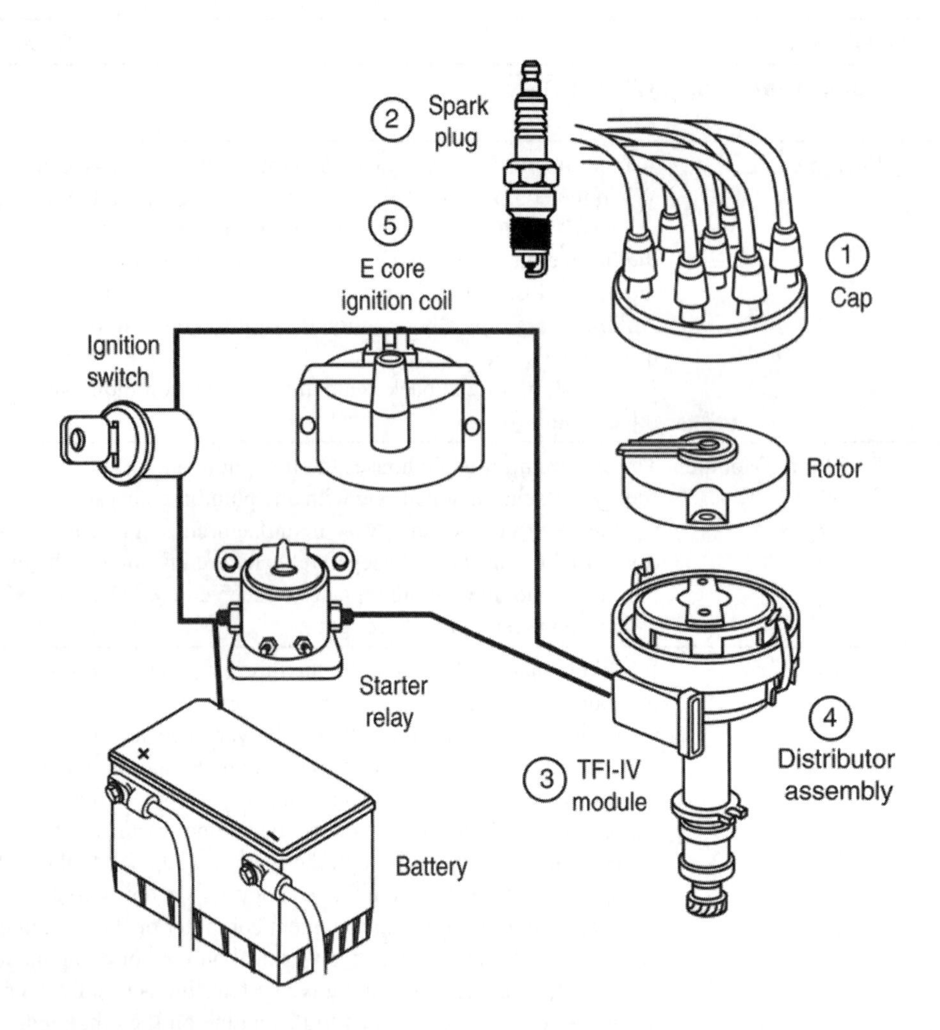

ELECTRONIC IGNITION SYSTEM WITH DISTRIBUTOR

Diagram Number	Component Name	What It Does
1.	Distributor Cap and Rotor	Probably the most familiar component with exception to the spark plug, the distributor cap has a small button in the center that carries current from ignition coil to the center electrode of the rotor. The rotor is attached to a shaft in the distributor that turns at the same speed as the engine's valvetrain. When the tip of the rotor passes under one of the contacts in the distributor cap, the electricity it is carrying finds a ground through the spark plug at the other end of an ignition wire that is attached to the plug and the cap. Due to the very high voltage levels that jump across these air capped gapped connections, the cap and rotor are subject to deterioration over time.

ELECTRONIC IGNITION SYSTEM WITH DISTRIBUTOR		***(Continued)***
Diagram Number	**Component Name**	**What It Does**
2.	Spark Plug	Spark plugs are very simple devices. They provide a ground for the ignition system through the threaded body that bolts into the cylinder head. When current is applied to them, a small lightning bolt jumps across the inner electrode that is insulated from the outer shell to the outer electrode attached to the spark plug shell. This spark starts a chemical reaction in the combustion chamber that causes the fuel and air mixture to oxidize very rapidly, heating it and causing expansion. Spark plugs may have electrodes made of, or coated with, copper, platinum, or titanium to help them last longer. When spark plugs wear the electrodes that are very square to the eye start to round and erode. This wear will cause the ignition system to work harder to make that spark jump the gap.
3.	Ignition Module	The ignition module receives a low-voltage pulsed signal from either the distributor pick-up coil, the PCM, or the crankshaft position sensor. The ignition module then amplifies the pulsed signal from 5 volts, or less, to a 12-volt signal that is sent to the coil(s). The ignition module amplifies the signal again to several thousand volts firing the spark plugs. In computer-controlled applications, the ignition module usually acts as middle management between the PCM and the ignition coil. In this role it takes commands from the PCM for timing and often provides the fixed or "base" timing necessary to start the engine. It then keeps the engine running until the computer takes over the responsibility.
4.	Distributor	The distributor has a mechanical connection to the engine and turns at the speed of the valvetrain, one-half crankshaft speed. In earlier vehicles that had a mechanical spark advance, the distributor had a vacuum controlled advance mechanism as well as centrifugal advance. Modern vehicles that are equipped with distributors use computer-controlled timing.
5.	Ignition Coil	The ignition coil takes a low voltage pulsing signal from the ignition module and amplifies it from about 9–12 volts to 5,000–40,000 volts and sends it to the spark plugs. Coils may be mounted remotely using an attaching high-tension wire. They may be integral or directly mounted to the spark plug, referred to as a “COP” (Coil Over Plug) design.

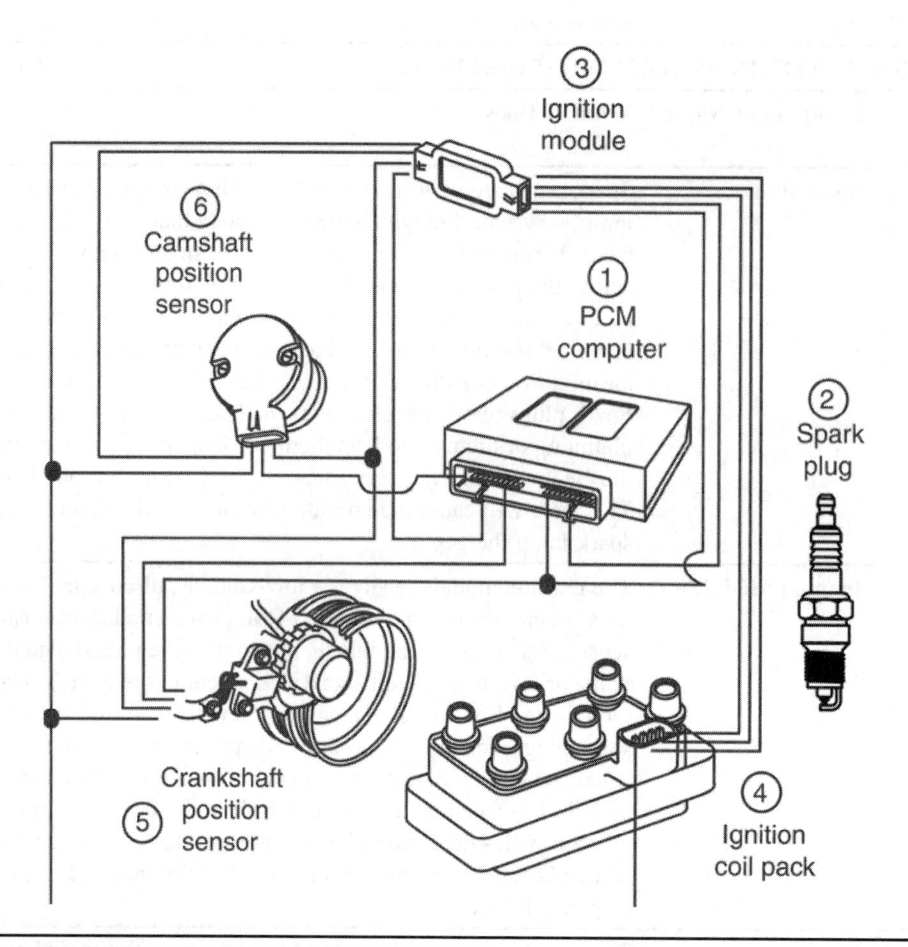

ELECTRONIC DISTRIBUTORLESS IGNITION SYSTEM

Diagram Number	Component Name	What It Does
1.	PCM (Computer)	The power train control module, or PCM, is the engine control unit. The PCM takes inputs from various sensors and operates various actuators. here is an extensive description of PCM operation in the emission section. For the purposes of ignition control, the PCM manages ignition timing by varying a signal between the ignition control module and itself. If the vehicle is equipped with knock sensors it will control timing when the engine experiences spark knock or pinging, usually by retarding timing.
2.	Spark Plug	Spark plugs are very simple devices. They provide a ground for the ignition system through the threaded body that bolts into the cylinder head. When current is applied to them, a small lightning bolt jumps across the inner electrode that is insulated from the outer shell, to the outer electrode attached to the spark plug shell. This spark starts a chemical reaction in the combustion chamber that causes the fuel and air mixture to oxidize very rapidly, heating it and causing expansion. Spark plugs may have electrodes made of or coated with copper, platinum or titanium to help them last longer. When spark plugs wear the electrodes that are very square to the eye start to round and erode. This wear will cause the ignition system to work harder to make that spark jump the gap.

ELECTRONIC DISTRIBUTORLESS IGNITION SYSTEM		*(Continued)*
Diagram Number	**Component Name**	**What It Does**
3.	Ignition Module	The ignition module receives a low-voltage pulsed signal from either the distributor pick-up coil, the PCM, or the crankshaft position sensor. The ignition module then amplifies the pulsed signal from 5 volts, or less, to a 12-volt signal that is sent to the coil(s). The ignition module amplifies the signal again to several thousand volts firing the spark plugs. In computer-controlled applications, the ignition module usually acts as middle management between the PCM and the ignition coil. In this role it takes commands from the PCM for timing and often provides the fixed or "base" timing necessary to start the engine. It then keeps the engine running until the computer takes over the responsibility.
4.	Ignition Coil Pack	The ignition coil takes a low voltage pulsing signal from the ignition module and amplifies it from about 9–12 volts to 5,000–40,000 volts and sends it to the spark plugs. In distributorless ignition systems there may be multiple coils. In some applications cylinders share a coil, while coil on plug systems have a coil for each spark plug. These connect directly to the spark plug, and are referred to as a “COP” design.
5.	Crank Position Sensor	In the distributorless system the PCM or computer must have a way to determine what position the crank is in rotation so that it knows when to fire the spark plugs. This is accomplished by using a degreed wheel on the crank or flywheel with a special signature that tells the PCM which location is the number one cylinder. This helps the PCM synchronize the spark plugs to the crankshaft position. Because the crank turns 2 revolutions for each revolution of the valvetrain, spark plugs are fired twice during a complete cycle; once during the power stroke and once during the exhaust stroke. This is referred to as waste spark design.
6.	Cam Position Sensor	In some systems an additional sensor called the cam position sensor takes the place of the special signature on the crank sensor wheel. This special signature is the reference to start the sequence of ignition firing in the correct order to provide a start. This is used to determine the exact position of the camshaft and number one cylinder. This added sensor is most commonly found in systems that use sequential multi-port fuel injection when the PCM needs to fire both the injector and the spark plugs at precise times in reference to crank position.

Exhaust System

Since it is pretty apparent that exhaust systems are the plumbing coming out of the engine, we are going to keep this section brief. We are going to introduce you to the nomenclature you will need to identify components.

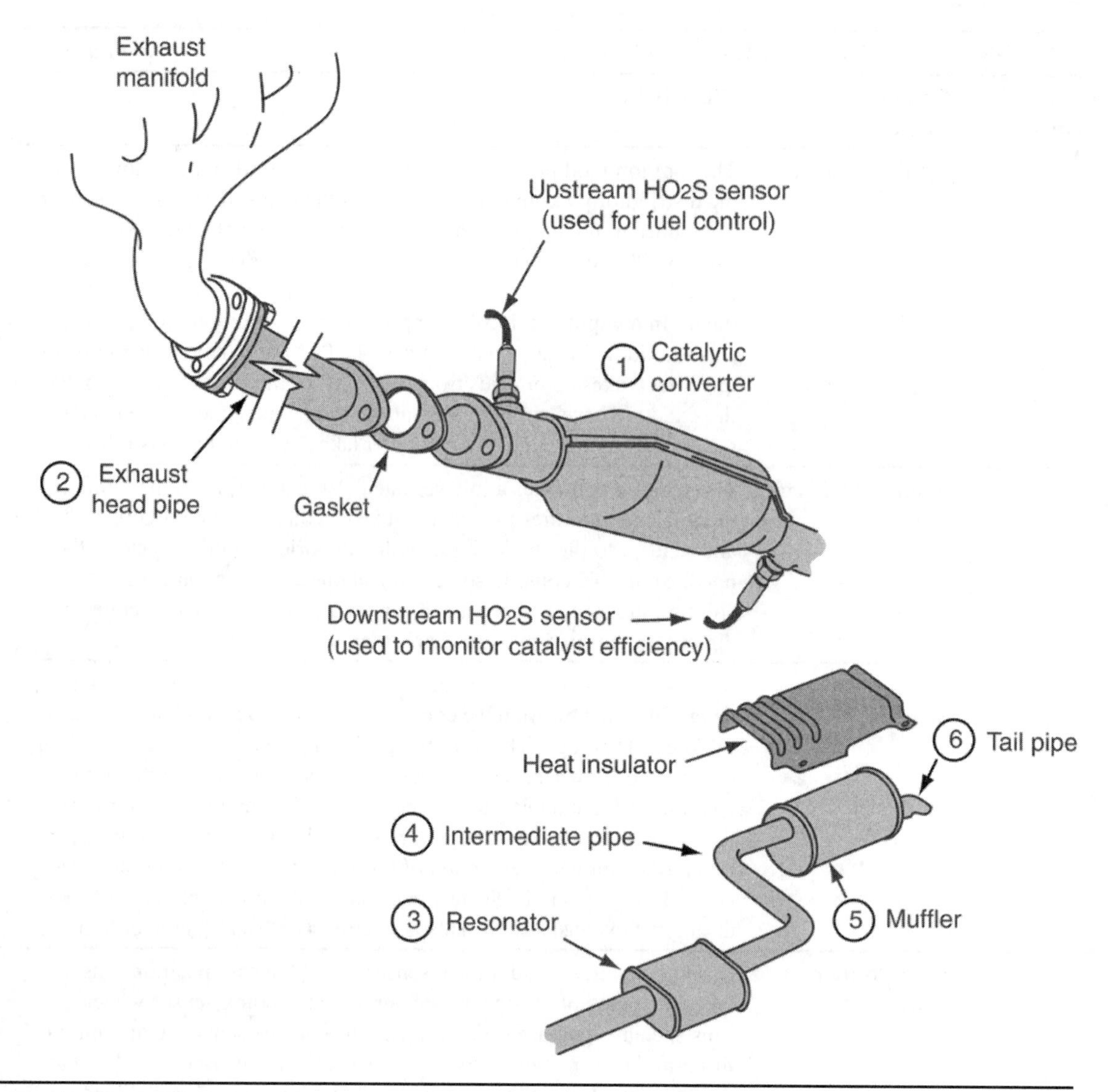

EXHAUST SYSTEMS

Diagram Number	Component Name	What It Does
1.	Catalytic Converter	The most complex component of the exhaust system is an emission component with no moving parts. The catalytic converter is located at the front of the vehicle usually as far forward as possible. The catalytic converter has the job of removing or reducing several byproducts of combustion. These components are carbon monoxide (an odorless, invisible, and deadly gas), Hydrocarbons (small quantities of unburned fuel) and NOx or oxides of Nitrogen, (not to be mistaken for nitrous oxide) which are responsible for photochemical pollution (acid rain). These emissions are removed by passing the exhaust stream over a bed of precious metals that cause a chemical reaction with the exhaust when they are hot. The conversion creates carbon dioxide and water in a perfect world. This is why you often see large amounts of water running out of the tailpipes of late-model vehicles and steam upon the warm-up of the engines.
2.	Head Pipe	The head pipe is the pipe directly behind the exhaust manifolds on the engine. It often has the catalytic converter built in.

EXHAUST SYSTEMS		*(Continued)*
Diagram Number	**Component Name**	**What It Does**
3.	Resonator	Some exhaust systems have a resonator in them to help remove some of the engine's noise. These may be located before or after the muffler depending on what the engineers were trying to accomplish. You can think of it as a pre or post-muffler muffler.
4.	Intermediate Pipe	The pipes between the head pipe and the muffler are known as intermediate pipes. On V-engines with dual exhaust sometimes there is a pipe that joins the two sides of the engine in the intermediate pipe. This pipe is called a crossover or H-pipe.
5.	Muffler	The muffler is a resonating chamber that cancels out loud noises from the engine. There is usually a perforated tube that runs through the muffler. The perforations allow sound to be reflected to the chambers within the muffler. These are sometimes filled with damping material like steel wool or ceramic panels.
6.	Tailpipe	Just like it sounds, this is the pipe that exits the exhaust out from under the vehicle at the rear or side of the vehicle.
7.	Oxygen Sensors	These sensors monitor gases passing through the exhaust piping. They provide critical input for emission control.

Emission Control Systems

Emission control systems encompass nearly every engine control system. We have discussed the evaporative system in the fuel system section and the catalytic converter in the exhaust. Round out your knowledge of the basics of emission control systems by looking at a general computer-controlled system and some of its possible emission control devices.

Because different vehicles require different types of equipment, you will not find really specific system questions here, but you will greatly benefit from understanding how the systems work. These can be very expensive components to repair and they seldom have much effect on the way the vehicle runs. This is why the service consultant must be able to help the customer understand the implications of driving a vehicle with the malfunction indicator light (MIL) on. The MIL lamp is amber in color to indicate caution.

The modern engine-management system is an amazingly adaptable and dependable system. It makes vehicles start and run much better and more consistently than pre-computer-controlled vehicles. It might surprise you to know that the original reason to have on-board computers was actually to control emissions. The easiest way to understand these systems is to relate them to our own bodies. The computer is like our brain. It has a vast store of information and "programming" in it. It can adapt to changes in weather, driving style, road conditions, and even vehicle wear. Unlike the humans who designed them, powertrain control modules (PCM) do not have to make many mistakes to learn and they can learn in milliseconds. That does not mean that when they get poor inputs they do not make bad decisions. When they do, the service consultant is the front-line psychologist to help the PCM get back on the right path.

So, how do these little boxes come to all these rapid and accurate decisions? One easy way to think of it is to imagine that they use electronics to replicate human senses. Start with smell: the computer uses one or more oxygen sensors mounted in the exhaust system to sniff the exhaust stream to see if it is delivering too much or too little fuel. It can even use this information to confirm a misfire condition or a bad catalytic converter. In reality, its sense of smell is just an electrical signal (usually .1 to .9 volts, or 100 to 900 millivolts) sent to the PCM and called an input. Moving on to feel or touch, the computer uses several sensors to this end. Coolant and air temperature sensors tell it the temperature. The mass airflow sensor feels the amount of air the engine is breathing and tells the PCM. A crank and cam sensor help the PCM see the engine's speed and track misfires. Many engines use a set of electronic ears to listen for pinging or detonation; thus this is labeled a detonation sensor. The PCM uses this information to adjust timing and protect the engine from damage. The PCM "sees" the position of the driver's foot on the throttle by tracking voltage by a variable resistor called a throttle position sensor. There are many other inputs to the PCM. They are included in the task list. Your shop's technicians can help you understand them when a diagnosis and repair require it.

The PCM takes all of these inputs and responds to provide the best emissions and drivability to the driver. This includes varying timing and fuel delivery many times per second; tracking inputs like air conditioner commands and power steering inputs to adjust idle speeds; and responding to load to vary the application of emission control devices like air pumps, evaporative systems, and EGR valves.

The EGR or exhaust gas recirculation system is used to reintroduce a controlled amount of exhaust back into the combustion process. Because exhaust is basically an inert gas that will not create any heat during combustion, it helps to cool the process down under light loads. This serves two purposes. First, it lowers NO_x emissions; and second, it helps to cool down the internal parts of the top end of the engine like cylinder heads, pistons, and valves, which can help to eliminate pinging that can occur under these leaner fuel conditions. Pinging or detonation occurs when the air and fuel mixture fires before the piston is all the way up on the compression stroke. Because the piston is connected to the crank, it must keep going up. However, the force created by the false ignition process makes a knocking or pinging sound as the flame front hits the top of the piston like a hammer. When this happens, the engine is misfiring and creating large amounts of all emissions. Prolonged detonation can severely damage the piston and/or engine.

Air pump systems have been in use since 1969 when some of the manufacturers used them to get their high-performance engines past the California emissions standards. This system pumps ambient air into the exhaust stream to help finish combustion of any burnable by-products in the exhaust system. Air injection is necessary for some types of catalytic converters to be efficient. The system is made up of a belt-driven or electric pump on the engine, lots of plumbing, and some one-way check valves to keep exhaust from getting into the pump. These systems are controlled by systems ranging from simple vacuum valves to complex strategies built between the PCM and the air management valve. These systems allowed manufacturers to continue to use older engine designs that were not as efficient and maintain lower emission standards. Most vehicles no longer use air pumps, as newer engine designs have found more efficient means to lower emissions.

Cranking (Starting) and Charging Systems

This section will not have charts because you are not expected to know about the inner components of the cranking or charging systems. You are going to need to know what the components are and how they work. Start with the cranking system.

When drivers wants to start the engine, they begin by turning the key that sends power to a high-amp relay. The relay may be an actual relay or it may be a solenoid mounted on the fender or the starter itself. Because starting the vehicle takes 120 amps or more, this relay is needed so that the cranking current does not have to pass through the ignition switch. The relay is an electrical switch that connects the battery directly to the starter while it is engaged. The starting circuits are the highest amperage circuits in the vehicle and may be easily identified by their large cables. These include the battery cables. When all of the electrical connections are made, the starter (cranking motor) begins to spin, pushing a starter drive gear into the flywheel drive gear at the same time. Once the engine begins to run, the starter drive backs away from the flywheel because the engine is running faster than the drive. The driver releases the key and the starting system becomes dormant until the next start-up. Always remember, a cranking motor, many times referred to as a starter, only provides rotation for starting. Other requirements are needed for an engine to run. The most common causes of starter failure are failure to engage due to open internal circuits or high resistance between circuit connections.

The battery is simply a storage device. It stores the necessary electrical energy to start the vehicle and it provides a reserve when loads exceed the alternator's output. Battery cables are the most frequently overlooked cause of repeat failure of alternators and batteries.

The alternator (in the past and recently referred to again as the generator) and voltage regulator are the charging system components. Most modern alternators have integral voltage regulators, which means that the voltage regulator is inside the alternator.

Automotive electrical systems are direct current or DC voltage. Alternators perform their work in alternating current or AC voltage. This AC voltage is converted to DC voltage inside the alternator through the use of diodes. Then the voltage is controlled by the voltage regulator. According to Ohm's law, the more load that is on an alternator, the less voltage it will produce. The opposite is also true. The voltage regulator controls voltage, attempting to maintain around 14.3 volts. Alternators are generally serviced as complete units, so the cause of failure is often boiled down to inability to meet specifications for amperage loads or minimum voltage. Alternators are very hard-working components in late-model vehicles. Some are capable of generating as much as 130 amps. Failure occurs due to wear and heat in most cases.

2. Tasks 1, 2, and 3 Drivetrain Systems (4 Questions) (includes manual transmission/transaxles, automatic transmission/transaxles, and drivetrain components)

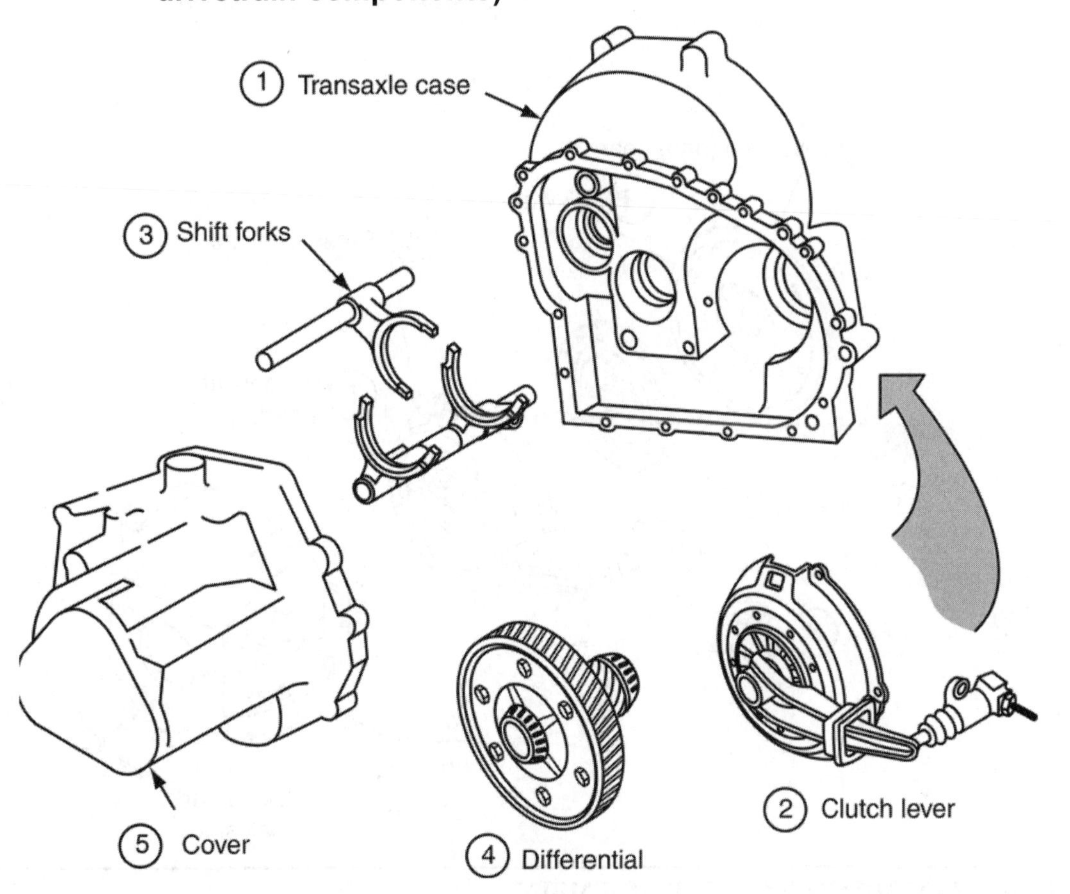

Manual Transmissions/Transaxles

The manual transmission uses a clutch and manual shifting to change gears. We will look at a basic transaxle to identify the components. You do not need to make diagnostic decisions in this test. Diagnosis should always be left to the technician. You only need to know how the system functions and how to identify major components.

MANUAL TRANSMISSION/TRANSAXLE

Diagram Number	Component Name	What It Does
1.	Transaxle Case	All of the components attach or are housed by this assembly
2.	Clutch Lever	The external connection to either the hydraulic cylinder or cable that operates the clutch inside the bell housing.
3.	Shift Forks	Manual transmissions use forks that straddle the synchro/gear assemblies to move the synchro from gear to gear causing a shift.

MANUAL TRANSMISSION/TRANSAXLE		***(Continued)***
Diagram Number	**Component Name**	**What It Does**
4.	Differential	In a front-wheel drive, the differential is usually inside the case and driven by a final drive gear from the transmission gear train that engages the large ring gear.
5.	Cover	An access point is available on some transaxles. The final drive gears are often in this cover.

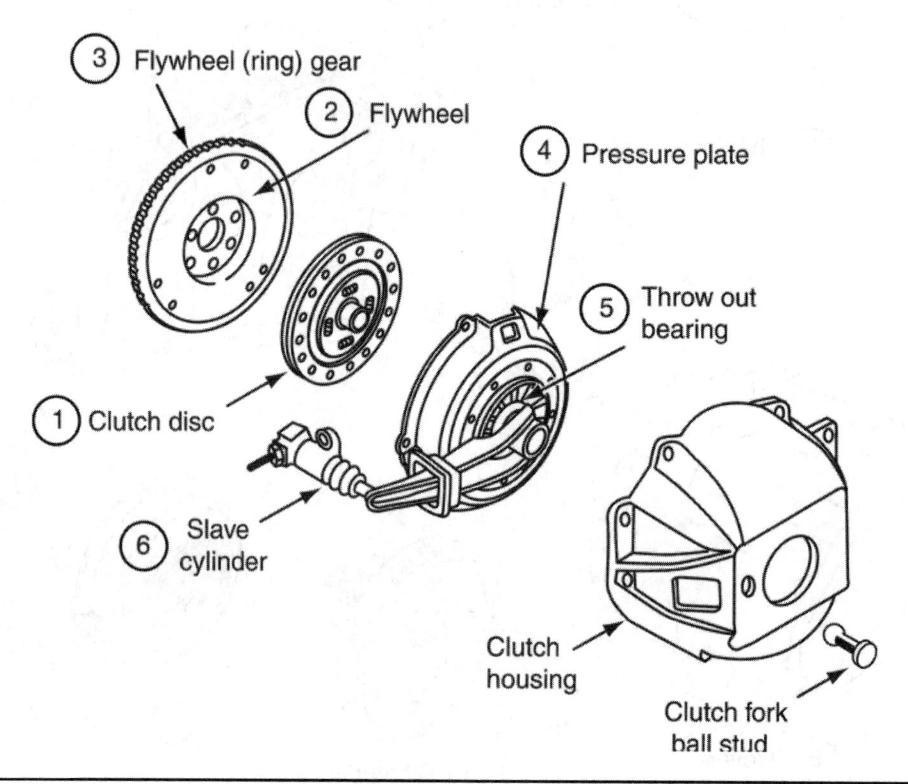

MANUAL TRANSMISSION: CLUTCH ASSEMBLY		
Diagram Number	**Component Name**	**What It Does**
1.	Clutch Disc	The friction part of the clutch. The springs in the center are there to cushion the clutch engagement. The clutch disc is two sided and riveted together. The disc slides onto and drives the transmission input shaft.
2.	Flywheel	Provides one side of the disc-clamping surface. The flywheel is machined smooth before assembly and bolts to the rear of the crankshaft.
3.	Flywheel (Ring) Gear	This gear is pressed on to the flywheel and is the gear the starter engages when turning the engine over during starting.
4.	Pressure Plate	Plate Provides the clamping force and is actuated by the clutch fork to release the clutch.

MANUAL TRANSMISSION: CLUTCH ASSEMBLY		*(Continued)*
Diagram Number	**Component Name**	**What It Does**
5.	Throw Out Bearing	This bearing is a sealed bearing that rides against the pressure plate actuating fingers during disengagement. Many late model vehicles apply light pressure so that the throw out bearing is in constant contact with the pressure plate.
6.	Slave Cylinder	Some clutches use a cable to connect the pedal assembly to the clutch actuation fork. Others use a hydraulic system similar to brakes. The slave cylinder may be inside or outside the bell housing. The slave cylinder expands to push on the clutch fork when the pedal is depressed and the clutch master cylinder generates hydraulic pressure. This action and relationship is similar to a brake master and wheel cylinder.

Automatic Transmissions/Transaxles

The modern automatic transmission is a fairly complex unit. Just like the manual transmission, it provides different gearing to help the engine move the vehicle. Unlike the manual, it uses a series of bands that grab hold of spinning drums and a planetary gear set that can create several different gear ratios depending on how it is applied by the transmission. The automatic transmission is controlled hydraulically. Instead of a clutch that must be engaged and disengaged, it uses a hydraulic coupling device known as a torque converter that varies the amount of engine torque and engagement. The internal controls of the transmission are fed hydraulic pressure by pumps inside the transmission and controlled by the valve body that directs shifts by changing pressures and directing fluid through passages. There is another style Automatic transmission out that is called continuously variable transmission or CVT that has less gears and is smoother shifting than the traditional automatic transmission since it has no shift points and the transmission is the transmission of choice in the newer hybrids. Below is a diagram of some of the major components of the automatic transmission, along with a chart of their names and purposes. You will see many similar components with our manual chart.

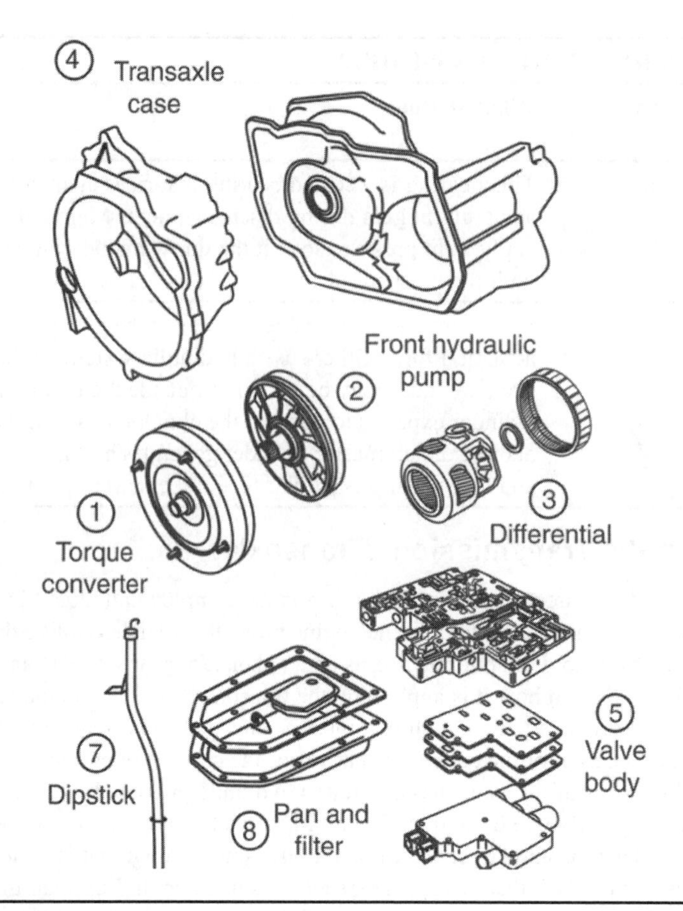

AUTOMATIC TRANSMISSION/TRANSAXLE

Diagram Number	Component Name	What It Does
1.	Torque Converter	A hydraulic coupling device that connects the engine to the transmission. Most late model automatics have an electric locking clutch inside the converter that makes a direct connection between the engine and the transmission during light load cruising. This drops rpm and helps with gas mileage. This is referred to as a lockup torque converter. This design now raises the miles per gallon (MPG) used by an engine with an automatic transmission to be the same or similar to one with a manual transmission.
2.	Front Hydraulic Pump	This is one of the pumps in a transmission. It is the only pump in this transmission. Some transmissions have a rear pump as well.
3.	Differential	In a front-wheel drive the differential is usually inside the case and driven by a final drive gear from the transmission gear train that engages the large ring gear.
4.	Transaxle Case	All of the components attach to or are housed by this assembly.
5.	Valve Body	The hydraulic control unit. Newer electronic transmissions have solenoids within the valve body that are commanded by the PCM.
Not Pictured	Side Cover	Usually a cover for the valve body in front-wheel drive applications. Most rear-wheel drive vehicles have their valve body in the bottom of the transmission in the oil pan area.
7.	Oil Dip Stick	Most automatic transmissions have a fluid level dip stick, while most manual gearboxes do not.

AUTOMATIC TRANSMISSION/TRANSAXLE		*(Continued)*
Diagram Number	**Component Name**	**What It Does**
8.	Pan and Filter	Older automatic transmissions use filters that can be replaced as maintenance service. Most vehicles are equipped with a screen that does not require maintenance. These vehicles simply require fluid changes.
Not Pictured	Planetary	The planetary is a multiple-ratio gear set used in automatic transmissions to provide different gear ratios in a compact package. It's called a planetary because it is a system of a single gear with multiple gears that revolve around it, similar to the way the planets revolve around the sun.

Drivetrain Components

Drivetrain components include the rest of the components that connect the engine and transmission to the wheels. The components include: drive shafts, (CV) axles, differentials, and four-wheel drive or all-wheel drive transfer cases. There are so many designs, we are going to take a look at generic versions of each to help you understand how they work and interrelate.

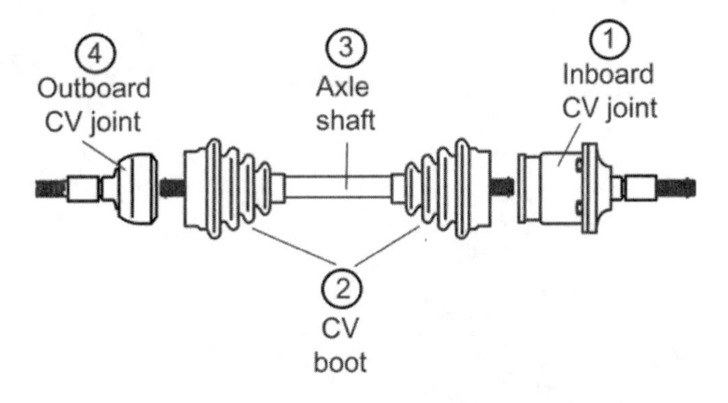

DRIVE SHAFTS: CV AXLE		
Diagram Number	**Component Name**	**What It Does**
1.	Inner Constant Velocity (CV) Joint	The Inner joint snaps into the differential side gears. This joint only operates in two planes. It must handle in and out movement called plunge as the vehicle moves up and down changing the length of the driveshaft. It must also handle moving up and down with the suspension.
2.	CV Boot	The major maintenance item on CV axles. The outer boots have the highest failure rate due to all the movement they must provide. The boot holds the lubricant in the joint and has bellows-shaped ridges that help it to move with changes in position and to keep contaminates like dirt, water, etc. out.

DRIVE SHAFTS: CV AXLE *(Continued)*

Diagram Number	Component Name	What It Does
3.	Axle Shaft	The axle shaft has splined areas at each end to engage and mount both CV joints. It may be hollow or solid steel.
4.	Outer CV Joint	The workhorse of the CV axle, this joint must operate in multiple planes at one time while rotating. The outer joint must turn with the wheels and move up and down with suspension. You will find CV axles used in front-wheel drive vehicles to both wheels, on front axles in late model 4-wheel drives with independent suspension, rear differential applications on mini 4-wheel drives, and even on some drive shaft applications.

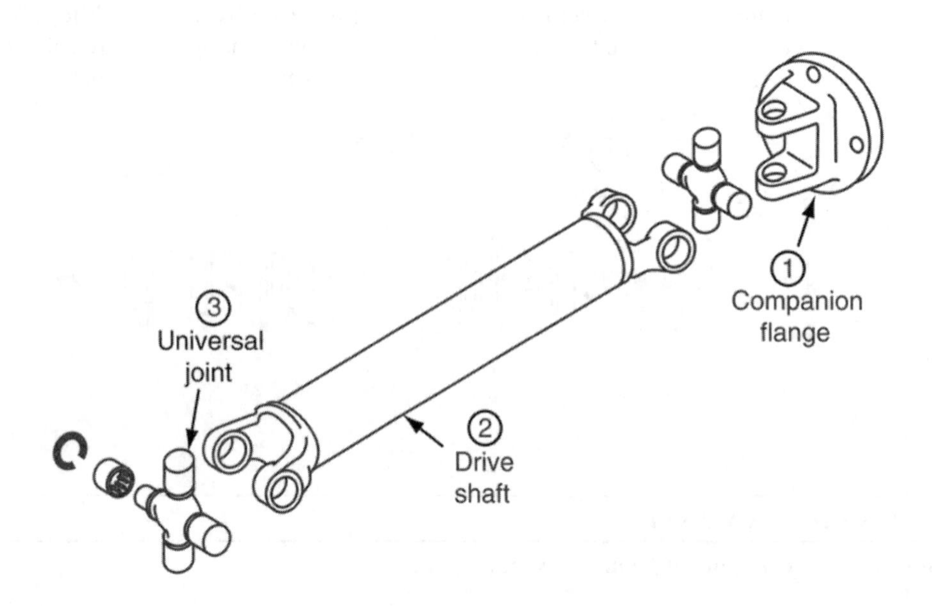

DRIVE SHAFTS: CONVENTIONAL

Diagram Number	Component Name	What It Does
1.	Companion Flange	This type of flange bolts to a differential or transfer case flange. Many vehicles will place the actual U-joint bearing cups directly into a yoke on the differential or transfer case output and retain them with a U-bolt. This is becoming the standard approach because of the repeatability of installation, which helps to control vibration.

DRIVE SHAFTS: CONVENTIONAL		*(Continued)*
Diagram Number	**Component Name**	**What It Does**
2.	Driveshaft	This component provides the connection between the transmission or transfer case and the differential. They may be constructed of solid or hollow steel tubing, aluminum, or carbon fiber. The U-joints are pressed into the drive shaft and retained by clips in most applications.
3.	U-joints	U-joints allow the drive shaft to make a connection while rotating (universal joints) and moving up and down with the suspension. There are four cups that hold needle bearings that allow the U-joints to turn to accommodate the movement.

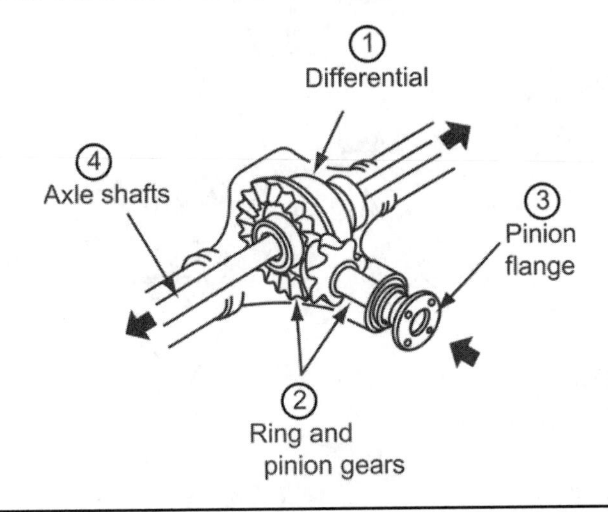

DIFFERENTIAL		
Diagram Number	**Component Name**	**What It Does**
1.	Differential	Differentials may be found at either the front or rear of the vehicle. Front-wheel drive vehicles incorporate the differential into the transaxle. 4-wheel drives use a differential at each end of the vehicle and rear-wheel drives have one at the rear of the vehicle. It is a safe conclusion that the name that precedes drive tells you where differentials are located. So what does the differential do? In our picture you will see that the arrow points to a component. This is the carrier and it contains the actual differential. Differential allows the drive wheels to turn at different speeds when the vehicle is turning a corner. In this situation the outside wheel has a greater distance to travel than the inner wheel. If they were locked together it would be a pretty jerky ride if the wheels could not travel at different speeds. Limited slip differentials use clutches to lock the two wheels together, when going in a straight line, that slip to help to make turns.
2.	Ring and Pinion Gears	In a conventional rear-wheel or 4-wheel drive vehicle, the driveshaft connects to one of two large gears called the pinion gear. The gear that it meshes with changes the direction of the driveshaft rotation 90 degrees. It is called the ring gear. The ring gear is bolted to the carrier and causes the carrier to turn. In front-wheel drives the engines crank runs parallel to the drive axles. There is a form of final drive from the transmission part of the transaxle that drives the ring gear and carrier assembly.

DIFFERENTIAL		***(Continued)***
Diagram Number	**Component Name**	**What It Does**
3.	Pinion Flange	This provides a means to connect or couple the driving shaft to the rear assembly.
4.	Axle Shafts	These shafts, usually solid steel, rotate at different speeds and provide connection to the wheel and tires.

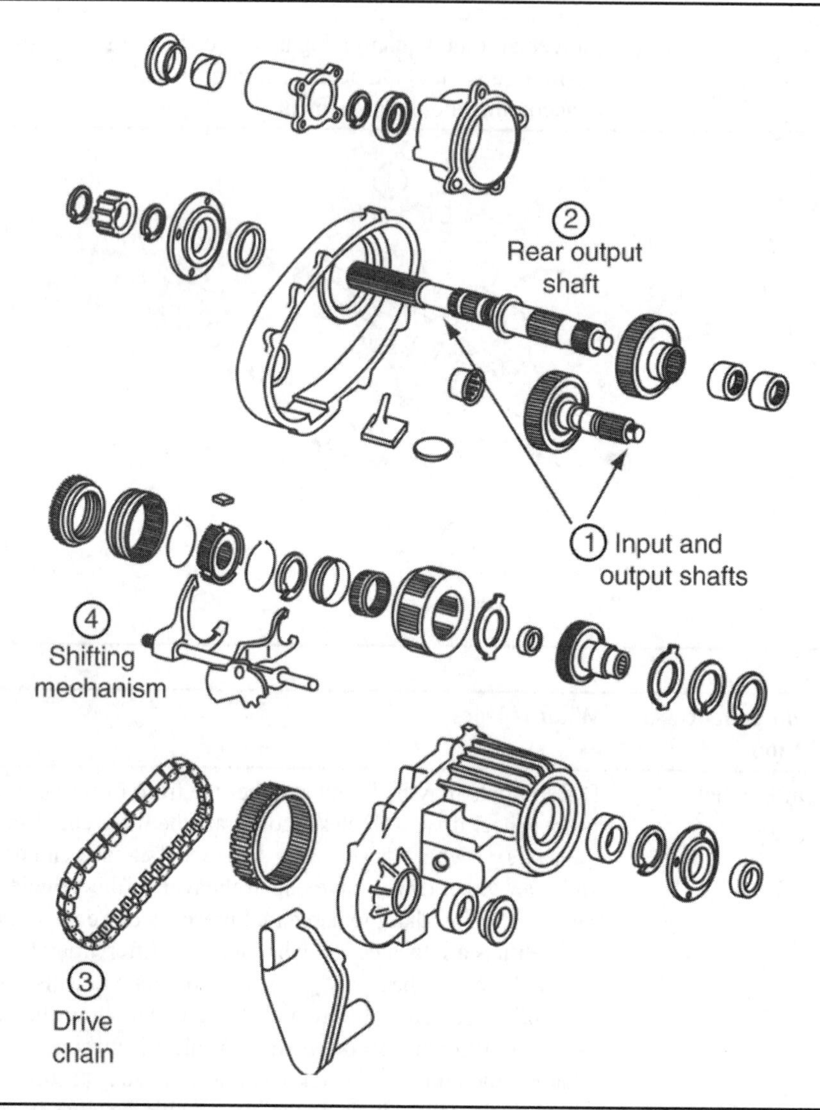

4-WHEELDRIVE TRANSFER CASE

Diagram Number	**Component Name**	**What It Does**
1.	Input and Front Output Shafts	We want to draw your attention to just four areas and features of a transfer case. First let's discuss what a transfer case does. Number one identifies the input and output shaft areas. A transfer case is a gearbox that engages or disengages the front and rear axles. There are usually two different ratios inside the box, which we will discuss more below. The transfer case is either mounted behind or physically bolted to the rear of the transmission via the input shaft. The front output shaft connects to the front driveshaft and the front axle/differential.

4-WHEELDRIVE TRANSFER CASE		*(Continued)*
Diagram Number	**Component Name**	**What It Does**
2.	Rear Output Shaft	The rear output connects to the rear driveshaft and rear axle/differential. This is the shaft that is used in both 2- and 4-wheel drive. Unless the transfer case is shifted to neutral, this drive shaft always is driven by the engine and transmission providing vehicle movement.
3.	Drive Chain	Many transfer cases use a chain to drive the front output shaft off of the main shaft when 4-wheel drive is engaged. Other transfer cases use a gear-to-gear front drive. This is very heavy duty in construction.
4.	Shifting Mechanism	Conventional transfer cases have a shift lever inside the vehicle to allow the drive to select between 2-wheel drive (rear wheels only), 4-wheel low (all 4 wheels driven slower while the engine runs faster. This helps to climb hills or do heavy work.), and 4-wheel high (all 4 wheels driven at normal speed) Some transfer cases have a synchronized shift to allow them to be shifted "on the fly." Newer systems have electronics that use a motor to change the gear selection instead of a shifter. There are many variations, but these are the most common. You will be adequately prepared if you understand what the transfer case does.

3. Tasks 1, 2 and 3 Chassis Systems (4 Questions) (includes brakes, suspension, steering, wheels, and tires).

Brakes

We are going to spend some time looking at the brake system, because it is a system that service consultants must be well versed in. In the diagram and chart we are going to look at is a typical disc/drum system and a generic antilock brake system.

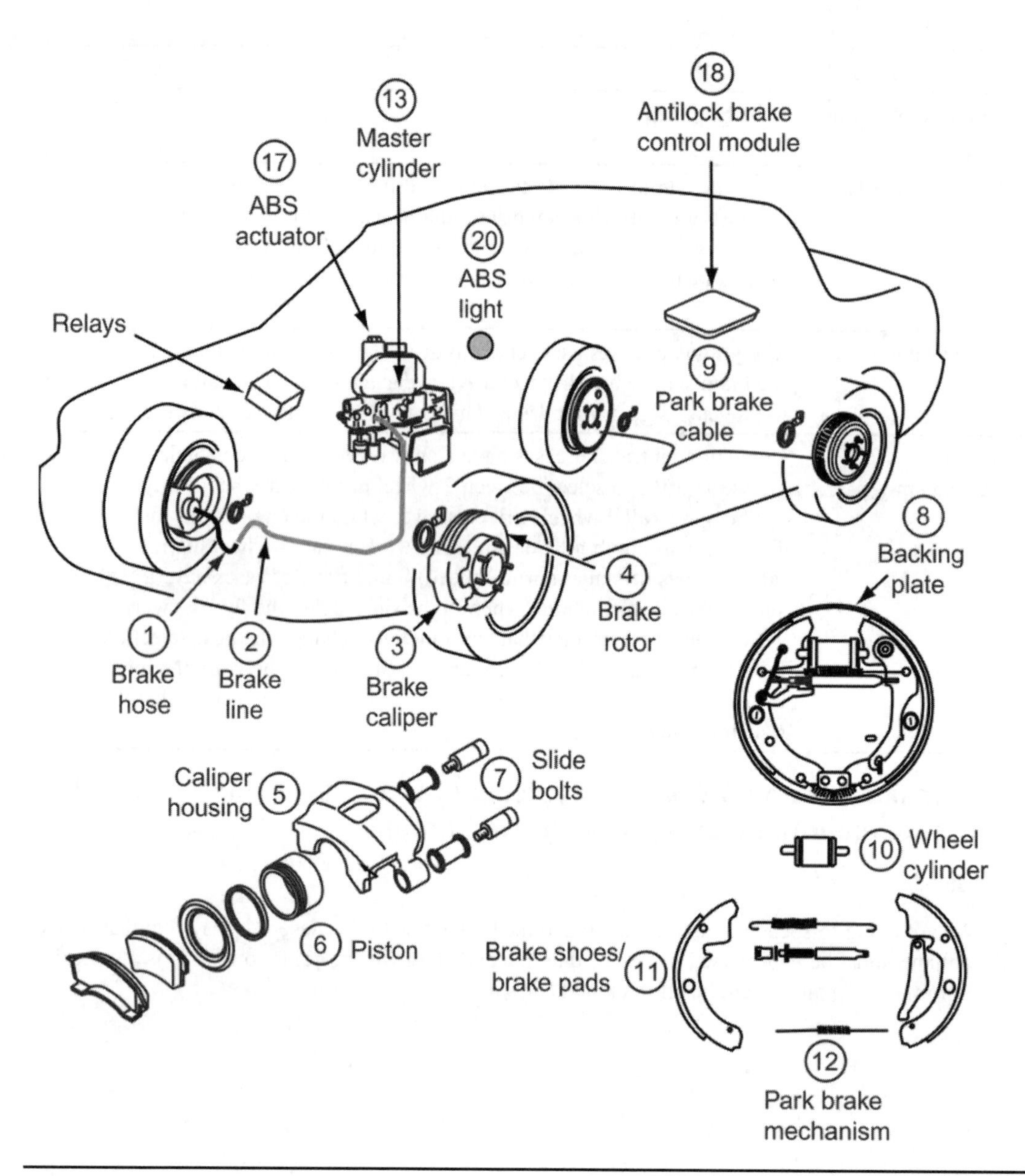

BRAKE SYSTEM

Diagram Number	Component Name	What It Does
1.	Brake Hose	Flexible hoses that connect the brake components at the wheels to the hard brake lines on the body. They allow suspension movement.
2.	Brake Line	Nonflexible hard lines that are attached to the body of the vehicle and run from the various components starting at the master cylinder and working out to where the brake hoses connect to them.
3.	Brake Caliper	The brake caliper is the moving component of a disc brake. It works like a clamp to grab the spinning rotor and slow the vehicle.

BRAKE SYSTEM ***(Continued)***

Diagram Number	Component Name	What It Does
4.	Brake Rotor	This is the namesake of the disc brake system. The rotor has two machined sides that allow the caliper to squeeze. The rotor rotates at wheel speed and also acts as a heat sink to shed high temperatures generated by friction due to stopping demands.
5.	Caliper Housing	This is the bare housing component of the caliper that functions as a mounting bracket. The piston bore is part of the housing and often has machined surfaces that allow the assembly to "float" on pins or slides to compensate for wear of the brake pads.
6.	Caliper Piston	The piston is more like a deep round puck. It may be made of steel, aluminum, or special plastic compounds. Most brake calipers have only one; but high performance applications and heavy vehicles may have multiple pistons. When the brake pedal is depressed the master cylinder creates pressure in the system which causes the piston to push out of its bore toward the brake rotor the opposing force of the brake pad on the opposite side of the rotor causes a clamping force that stops the vehicle.
7.	Slide Bolts	Many brake systems use bolts that tighten into the spindle. The caliper floats on these. Others may use plates that allow the caliper to slide. These slides are critical to even brake pad wear. If the caliper does not slide, the inner pad often wears out very quickly because the pressure on the rotor is not equal on both sides.
8.	Backing Plate	On drum brake systems the backing plate is the mounting plate for all of the components of the system.
9.	Park Brake Cable	If you follow the diagram you will see that the park brake cable connects at one end to the parking brake pedal or lever and at the other end it attaches to the drum brake by a special cam mechanism that pushes the brake shoes out mechanically to secure the vehicle in a parked position. Rear disc applications use a piston-locking mechanism to hold the brake in park.
10.	Wheel Cylinder	The wheel cylinder performs the same purpose as the brake caliper. Instead of clamping though, it pushes out on the brake shoes.

BRAKE SYSTEM		***(Continued)***
Diagram Number	**Component Name**	**What It Does**
11.	Brake Shoes/Brake Pads	The component that is designed to wear out as it does its job, the brake shoe or disc pad has a very complex job. These parts are made up of metal and organic materials that are called friction materials. This is easier to understand if you understand how brakes do their work. Let's agree that there are two types of stops; the slowing stop and the panic stop. The brake pad takes the motion energy of the vehicle and converts it to heat. The rotor and drum provide a smooth surface for maximum heat transfer. The drums and rotors then radiate this heat out to the atmosphere to help keep the brake pads cool enough to do their job. Now back to our two kinds of stops. Brakes must do an amazing job of performing the slow stop, which is a progressive conversion of motion to heat, and then making the panic stop when called upon. This grace under pressure is not how brakes have always worked. Many older vehicles would lock up the brakes under any more than very gentle application. So, next time you stop your vehicle, take a second to feel how smoothly it can perform each type of function. Even under a panic stop, newer systems that use anti-lock controls provide safety in stopping under unusual conditions.
12.	Park Brake (Emergency Brake) Mechanism	If a hydraulic failure would occur, a backup mechanical system would be of help. The park brake mechanism mechanically actuates the rear brakes in all but a very few applications. When there are disc brakes on the rear, a similar device mechanically pushes the pads out to clamp the rotor. This handy safety device is almost always linked to the components that automatically adjust the brakes. So, using the park brake helps to adjust the rear brakes. Backing the vehicle up will also actuate most drum brakes self adjusters.
13.	Master Cylinder	This is the first component of the hydraulic portion of the brake system. When the brake pedal is depressed it generates pressure to actuate the brake calipers and wheel cylinders. Pressures can be as high as 2000 psi.
Not Pictured	Power Brake Booster	Brake boosters provide the assistance to push the brake pedal that is called power brakes. Most use vacuum applied to the master cylinder side of a chamber with a diaphragm in the center to supply this assist. Some systems are known as hydro-boost and use a similar arrangement with the power steering pump supplying hydraulic pressure for assist.

BRAKE SYSTEM		***(Continued)***
Diagram Number	**Component Name**	**What It Does**
Not Pictured	Pedal Assembly	The pedal is the component that attaches to the master cylinder pushrod and gives the driver a mechanical advantage due to a pedal ratio designed into the mechanism. This makes it so that even without assistance, 10 pounds of pedal pressure may create 100 pounds of brake pressure, similar to the advantage created by a lever.
Not Pictured	Combination/Proportioning Valve	In non-ABS vehicles this component is used to control the amount of pressure sent to the rear brakes. If pressures are equal at all four weight. Most vehicles are lighter in the rear than the front. Also note: when stopping weight transfer goes to the front brakes. This component usually has a pressure differential valve/switch that will shutdown part of the brake system should a large loss of pressure, like a leak, occur. It will also turn on the red brake light on the dash.
17.	ABS Actuator	An electro hydraulic component, this unit contains the electronic versions of the proportioning valves that are run by the ABS control module to control wheel lock up.
18.	ABS Control Module	This component analyzes inputs from the wheel speed sensors and sometimes the vehicle speed sensor to control wheel lock-up. This is very much a simplification because of the vast number of systems of varying complexity out there.
Not Pictured	Brake Warning Lamp	The red brake warning lamp comes on when either the park brake is not fully released, the brake fluid is low or the pressure differential switch in the combo valve has tripped. In some applications the ABS system can turn this light on along with its malfunction indicator light. Remember, red lights have higher significance than amber.
20.	ABS Warning Lamp	This light comes on when the ABS system is not functioning. Some systems turn it on when brake fluid is low. When both the ABS and the MIL lamps are on or flashing, it indicates immediate danger or component damage.

Suspension

The suspension is what holds the vehicle up, provides a smooth ride, keeps the tires in contact with the ground, and provides safe control when turning control. We are going to review each of the components in a short, long-arm (SLA) suspension, which is the most common type. The version we selected uses a suspension strut and is typical of both front and rear suspensions of many vehicles. Before we begin, let us quickly cover the two other types of suspensions commonly in use. The leaf spring suspension is found on the rear of many vehicles and the front of heavy-duty pickups. In this suspension, an axle housing is mounted with a leaf spring on each side of the vehicle. The spring functions as a control arm locating the rear end fore and aft in the vehicle in addition to providing normal spring function. More and more vehicles including trucks are being designed with independent suspension on all four wheels, so this design will probably only remain in heavy trucks in the future. The second variant is the MacPherson strut. In this design, the strut functions as the upper suspension attaching point. These suspensions do not have an upper control arm. The strut provides a pivot point at the top in the form of a strut bearing or mount that allowed the strut to turn with the wheels. This design is also giving way to much better handling than riding short, long-arm (SLA) designs. The strut pictured in the diagram is not a MacPherson strut because it is not a suspension member. It functions as the shock with an integral coil spring. The short, long-arm (SLA) suspension gets its name from the short upper control arm and long lower control arm that make up its design. This design handles very well because it allows the tire to move in the direction of negative camber as the body rolls going around corners. We will discuss camber when we look at alignment at the end of this section. These newer independent designs provide better tire-to-road contact. This not only promotes long tire wear but adds safety due to advanced tire friction-to-road contact or connection.

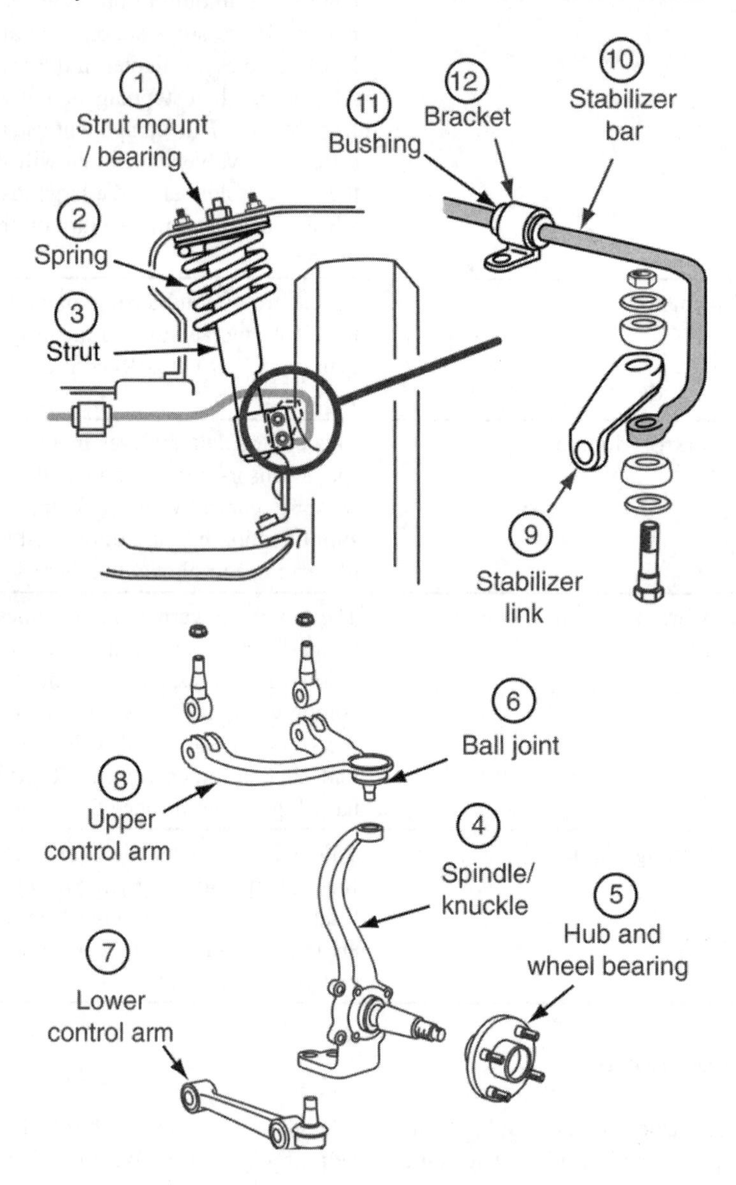

SUSPENSION

Diagram Number	Component Name	What It Does
1.	Strut Mount/Bearing	The strut mount cushions and isolates the strut assembly from the body. In McPherson applications it also houses a bearing to allow the strut body to rotate when the wheels turn.
2.	Spring	The coil spring is most common. The previously mentioned leaf spring is used with live axle applications, and one other variation known as the torsion bar round out the spring types used in modern vehicles. The torsion bar is a long straight spring that functions by twisting. It is attached at one end to the body and the other end to the suspension. It is usually adjustable to attain correct preload and ride height.
3.	Strut	The strut is a shock absorber that functions as a mount for the coil spring in some applications. The SLA suspension may use a strut like the diagram or it may have a shock absorber.
4.	Spindle/Knuckle	The steering knuckle or spindle, provides the movement during steering along with the mounting areas that accommodate the wheel hubs, brake components and the ball joint ends of the control arms.
5.	Hub and Wheel Bearing	This component is shown as a stand-alone piece but may be integrated into a brake rotor with serviceable bearings. The type shown uses a sealed bearing.
6.	Ball Joint	Ball joints are like the joints in our shoulders or hips. They provide for the multi-plane movement that occurs as the vehicle moves up and down and when the wheels turn. They get their name by a ball shaped bearing surface that rides in lined cup. The ball joint is usually bolted or pressed into the control arm and has a long stud that is bolted into the steering knuckle.
7.	Lower Control Arm	The lower arm is the longer one of the two. It joins the suspension Arm from the steering knuckle to the body of the vehicle. It, often has mounting points for shocks, stabilizer bars, and sometimes the springs.
8.	Upper Control Arm	The upper control arm is the short arm. It joins the suspension from the steering knuckle to the body of the vehicle. It, too, often has mounting points for shocks springs.
9.	Stabilizer Bar Link	There are several designs but the stabilizer bar link or end link connects the ends of the stabilizer bar to the suspension.
10.	Stabilizer Bar	Also known as a sway bar the stabilizer bar is an example of a torsion bar. When a force is applied to one side of it, there will be an equal and opposite force on the other end. This makes it so that as a vehicle rolls to the outside as it makes a turn, the force on the outside wheel will be transferred to the inside wheel to keep the tire in contact with the ground. This is sometimes referred to as an anti-roll bar.

SUSPENSION		***(Continued)***
Diagram Number	**Component Name**	**What It Does**
11.	Stabilizer Bar Bushing	To allow the stabilizer bar to pivot on the body of the vehicle, a bushing is located on the body equal distances from the center.
12.	Stabilizer Bar Bracket	This is the bracket that holds the sway bar and bushing to the body Bracket of the vehicle.

Steering

Only two types of steering are to be discussed; they are the rack and pinion and the linkage style. We will look at both of them. Some recent systems are electrically driven by motors using varying input voltage to a control module.

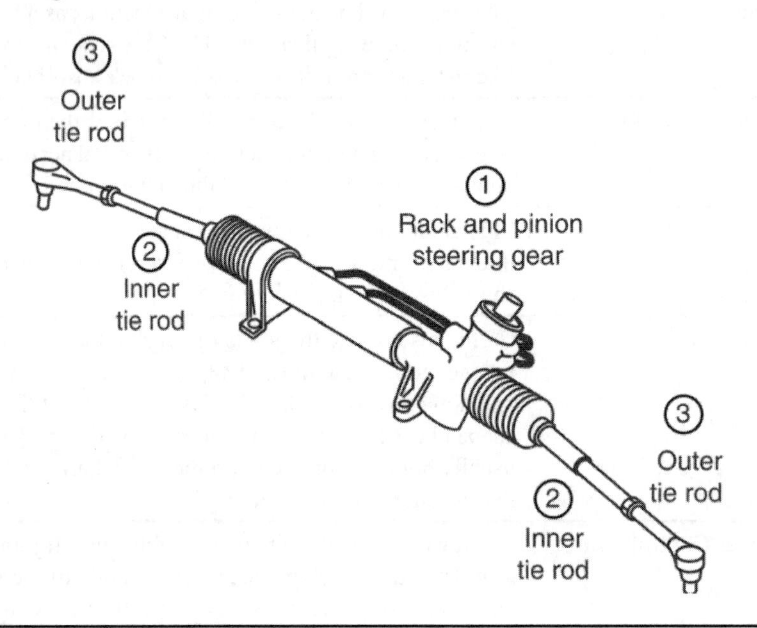

RACK AND PINION		
Diagram Number	**Component Name**	**What It Does**
1.	Rack and Pinion	Rack-and-pinion steering gets its name from the internal components. The rack is a long gear that runs horizontally inside the housing. The tie rods attach to the ends of the rack. The pinion is a small gear that connects to the steering column shaft. The reason for using gears is to change the ratio between the wheel's turn and steering input. For example, if you could turn the wheels all the way in one rotation of the steering wheel the vehicle would feel very twitchy and drivers would overcompensate in panic situations. These can be manual design, though most are power assist. When a rack and pinion is power assist a valve is located at the top of the pinion that provides hydraulic assist when there is movement of the pinion, and fluid lines will connect to the unit.

RACK AND PINION ***(Continued)***

Diagram Number	Component Name	What It Does
2.	Inner Tie Rod	Tie rods are like small versions of ball joints; they allow rotational and multi-plan movement to follow suspension movement. The inner socket on the rack and pinion is an open joint because a bellows boot similar to a CV joint boot covers it and protects it from contamination.
3.	Outer Tie Rod	The outer tie rod connects the inner tie rod to the steering knuckle. The two thread together allow alignment changes to the toe setting by adjusting their lengths. The use of left- and right-hand threads allows for this. Compare replacements with original for correct thread pitch and match.

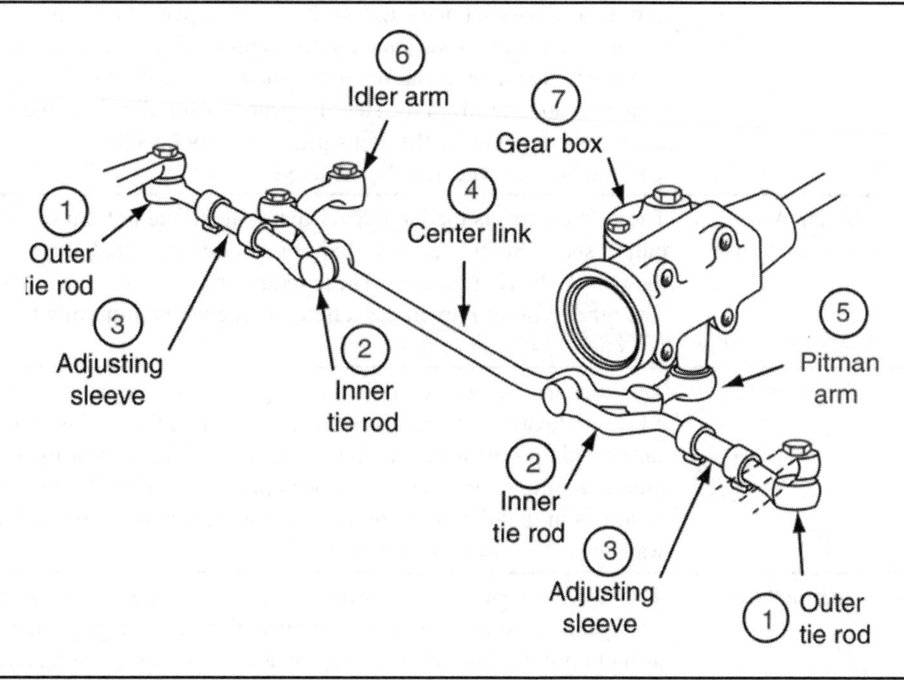

LINKAGE TYPE STEERING

Diagram Number	Component Name	What It Does
1.	Outer Tie Rod	Tie rods are like small versions of ball joints that allow rotational and multi-plane movement to follow suspension movement. The outer tie rod connects the inner tie rod to the steering knuckle. The two thread together and allow alignment changes to the toe setting by adjusting their lengths. The components of the suspension must move up and down at the wheels. The components of the steering must allow this movement while causing the least amount of change in the toe settings of the alignment. Tie rod ends allow around 30 to 60 degrees of movement. Looseness due to wear or binding are reasons for replacements.
2.	Inner Tie Rod	The inner tie rod works just like the outer tie rod. In most cases there is an adjusting sleeve between the two. The inner tie rods may attach to a center link, the tie rod for the other side of the vehicle, or directly to the steering box. Many inner and outer tie rods are threaded with one left and the other right handed to make it so that when the technician adjusts them they rotate in the same direction to adjust toe in or out on each side of the vehicle.

LINKAGE TYPE STEERING *(Continued)*

Diagram Number	Component Name	What It Does
3.	Adjusting Sleeve	The sleeves are generally made of steel and have clamps at each end where the tie rods thread into them. After the technician adjusts the toe to specification he tightens the clamps on the adjusting sleeve to lock the toe setting. When one or both tie rods are replaced the adjusting sleeves are often rusted and frozen necessitating their replacement. Often they are made with right and left handed threads so it is important to assemble and install them correctly.
4.	Center Link	The center link is almost always used in conjunction with an idler arm. The center link attaches between the steering box and idler arm and maintains a parallel plane across the front suspension. In most center link applications the tie rods attach to the center link and then to the steering knuckles. A similar variation of this theme is the drag link. The drag link generally attaches from the steering gear to a spindle. Like the center link, it does not provide for the adjustment but may be adjustable to allow the steering box, or wheel, to be centered.
5.	Pitman Arm	The Pitman arm is the large lever that bolts to the bottom of steering output shaft. Some may have ball joints like joints that attach them to the rest of steering linkage. These require replacement when they wear. The other type of Pitman has a hole to receive the joint of a tie rod and does not usually have to be replaced.
6.	Idler Arm	The idler arm is located on the passenger side of the vehicle. In all but a few applications, there is one idler arm. The idler arm is responsible for providing mirror image movement of the idler arm or its opposing pitman arm to keep the center link in precise parallel alignment. The bushings in the idler arm are the part that generally wear resulting in a wandering sensation on the road.
7.	Steering Gear	The steering gear may be power-assisted or manual. It has a gear arrangement using a set of ball bearings that run in a cage between the gears to provide smooth steering operation and minimum reaction back to the driver when going over bumps. This design is known as recirculating ball steering. This is the most prevalent design in use in the last 30 years. Common failures are usually leaks at output- or input-shaft seals and leaking power-steering line connections. The mechanical part of steering boxes is highly dependable due to the importance of its function. These systems are bulky and heavy compared to rack and pinion setups. They work best in heavy duty applications for these reasons.

Wheels and Tires

As with most of these areas, wheels and tires can be a specialty unto themselves.

Questions within this test will be limited to knowledge of tire and wheel service issues and not engineering. Look at common conditions that occur with tires. When a tire is out of round, it is called radial runout. If you spin a tire and wheel assembly on a balancer, you will nearly always see some radial runout. This will appear like high or low spots as the tire turns around. This can cause road vibrations if it is bad enough, even after multiple attempts to balance the assembly.

Another condition that occurs with defective tires, bent wheels, or tires that have had a blow to the sidewall is called lateral runout. Tires with this condition may appear to move side-to-side when spun on a balancer. The vehicle may demonstrate a vibration at speed and a sort of side-to-side wiggle when moving slowly.

Wheel balance and rotation are maintenance items for tires. Wheel balance is the process of spinning the wheel on the balancer. The balancer looks for the heaviest spot on the wheel and calculates the necessary weight to put on the spot directly across from it to equalize the assembly's dynamic weight. This is a severe over-simplification, but is more than enough understanding. The balancer uses both the inside and the outside of the wheel to work out a dynamic balance. The newest balancers apply road force to best provide solutions to the imbalance.

Wheels can contribute to these problems if they have been bent or are damaged from road hazards or improper installation. Most late-model steel wheels are somewhat flexible and can be damaged very easily. If the lug nuts that hold the wheels on are not properly tightened, a wheel can distort causing a problem that may feel like a tire problem.

Before we leave the chassis section, we need to take a look at alignment angles. The ability to explain the alignment angles to a customer is very valuable. Here are the basic angles that may show up in questions on the test.

The body appears to be going crooked down the road. Modern alignment equipment takes this angle into account when setting front toe to keep all four wheels going in the same direction. Readings are taken between all wheels through precise measuring instruments.

Camber

When you look at the wheel from the front, the amount the tire leans in or out at the top is called camber. When the tire leans out it is called positive camber and when it leans in at the top it is called negative camber. Camber is a wear angle; This means that if it is out of spec, it can cause the tires to wear on the shoulder to the direction the top of the tire leans. Out at the top causes outside shoulder wear; in at the top causes inside shoulder wear.

Caster

The best way to think of caster is to think of casters on the bottom of a chair. You can see that the post that goes up into the chair is ahead of the center of the wheel. Because of this, the wheel will always follow the chair's direction when it is pulled. When it turns, it will change direction to always go straight. If you try to turn the wheel so that it leads the post, it will quickly turn to follow again. This effect is called positive caster. If we were to move the post directly over the center of the wheel, it would not follow anymore particularly when turning. Caster is a not a wear angle. Excessive caster can cause a vehicle to steer heavily in parking lots or even make the steering wheel shake going over bumps. Vehicles with too little caster will wander and follow ruts in the road, causing an oversteer. In years past, cars without power-assist steering would be aligned with much negative caster to aid assist in turning at slow speeds.

Toe

Toe is the direction the front of the wheel points when going down the road. If you stand up and point your toes toward each other that is what is known as toe-in. If you point them away from each other that is known as toe-out. Vehicles with excessive toe-in will wear the tires across the face. They may not exhibit any unusual driving tendencies because toe-in is the more stable of the two settings. Vehicles with excessive toe-out will wear the tread across the face in the opposite direction. The vehicle may exhibit wander or be somewhat unstable when applying the brakes, produce squeal, or rapidly wear.

Thrust Angle

Thrust angle is the relationship between the overall toe direction of the rear wheels to the overall toe direction of the centerline of the chassis. If both rear wheels are pointed to one direction or the other, the front steering must compensate to make the car go straight. Vehicles that have a thrust line problem can often be spotted when following them. You can see the side of the vehicle from directly behind it, but not the other side.

Four-wheel Alignment/Two-wheel Alignment

Most late-model vehicles that are front-wheel drive and many rear-wheel drive vehicles require a four-wheel alignment. This means that sensors are mounted on all four wheels so that all of the wheels can be set. Traditional two-wheel alignment meant that only two sensors were put on the front of the vehicle with none on the rear. This method is not practiced anymore due to ability to use a thrust line on a rear-wheel drive vehicle. Even when only two wheels have adjustment, all four sensors are mounted to allow this compensation, and to provide a means to look for unforeseen damage or bent components.

4. 1, 2, and 3 Body Systems (4 Questions) (includes heating and air conditioning, electrical, restraint, and accessories)

Heating and Air Conditioning

The heating and air-conditioning systems provide comfortable air inside the cabin of the vehicle. Warm air is generated using the heat created by the engine. The conductor for this is known as the heater core. The heater core is a small radiator. Coolant from the engine is circulated through it and a blower blows across the core to provide hot air inside the vehicle. There are several passages for the air to flow through that are controlled by the controls on the dash. The controls may have cables, vacuum actuators, or electric motors that run a series of doors inside the ducts to direct the air where the driver wants it. Temperature is controlled by a blend door and in some cases a water control valve that limits the amount of warm engine coolant that enters the heater core.

Air conditioning works through the same set of ducts; but because there is no naturally cool fluid on a vehicle, it must create coldness. This is done by evaporation and condensation. The air-conditioning system uses a compound that has a very low boiling point. This refrigerant is simply called R-134A. The air-conditioning system takes advantage of the refrigerant's low boiling point to move it through two states, liquid and gas. Think of how water becomes steam when it boils and you will understand. Also, think of how a wet rag swung fast in the air becomes cold—this is called evaporation.

Air conditioners have a compressor that provides both suction and discharge. It creates movement and pressure within the system. The compressor pushes gaseous refrigerant to the condenser, which sits in front of the radiator. As the gas passes through the tubes of the condenser, it changes to a liquid and becomes colder. It changes state at some form of restriction which has suction applied. This liquid is then evaporated back to gas state. Only gas state refrigerant can pass through a compressor; liquid will not compress. Between the compressor and the evaporator there is a restriction. This can be in the form of a control valve or just a small tube called an orifice. This causes the liquid refrigerant to spray into the evaporator. At this point, the pressure on the liquid drops and it rapidly becomes a gas. This process is called evaporation. Think of our wet rag again. This causes the evaporator to become very cold so that a fan blowing across it will yield cold air. The compressor starts the whole cycle over again. The components of the air-conditioning system share most of the ducting with the heater but have electrical controls to engage the compressor. The system may have switches to protect the compressor from extremely low or high pressures. Many air-conditioning systems are monitored and controlled by a body control computer or the PCM.

Electrical

Current automotive electrical systems are 12-volt DC, and use the negative battery terminal as ground. There are many body systems that use electricity to operate. They include: power windows, door locks, cruise control, radios, lighting, seats, doors, antitheft systems, suspension controls, engine cooling fans, heating, air conditioning, etc. There is virtually no system on a modern car that is not controlled or monitored by an electrical device. This is why the condition and the capability of the charging system are so critical. The word electrical generally refers to electronic, that is, solid-state controls without moving parts, just transistors, diodes transformers, etc.

Restraints

Restraints refer to the safety items inside the vehicle. Seat belts and air bags are the most common items. Vehicles are equipped with all kinds of reminder and warning systems that deal with getting passengers to use the restraint systems.

The air bag or supplemental restraint system has a couple of key components of which you should be aware. The air bag modules themselves are the devices that rapidly deploy when a crash sensor detects that a blow to the vehicle might cause an injury that the air bag could lessen or prevent. These modules are for one-time use and must be replaced when they deploy. The other item that you should know about is the clock spring or spiral cable. It is most often damaged during steering service where the steering wheel is allowed to turn too many times without the steering box to limit it. This device is located under the air bag module inside the top of the steering column. It allows the steering wheel to turn roughly three full turns and provides an electrical connection to the air bag module and other components that may be in the steering wheel.

Seat belts are the other restraint area that may come up on the test. Modern seat belt designs allow the passenger to move their upper body freely to maneuver. They have a load-sensing inertia device in them that locks the belt under potential emergency or accident avoidance maneuvers. Some seat belts are equipped with pre-tensioners that fire off like air bags to keep the occupant in place during an accident. Child safety anchors were not included in the task list at this time.

Accessories

Accessories are our last and most broad area. Fortunately, you will not have to study this area much if you drive a car or truck. If you understand how mirrors, door locks, electric windows, and other normal body controls are used, you will do fine on any question asked here. Again, because every manufacturer has its own way of doing things, the questions have to be very general for the whole item-writing panel to agree to them.

Task B.5 Services/Maintenance Intervals (3 Questions)

Task B.5.1 Understand the elements of a maintenance procedure.

There are two different paths that can be taken when maintaining a vehicle. The first is following a maintenance schedule that performs standard items and services based on the wear characteristics of the vehicle. The second path is damage control repair upon failure. It has been proven in many studies that the maintenance path is the most cost-effective approach. To understand and explain a maintenance procedure, you must know what the procedure is trying to accomplish. Here is an example: Most manufacturers recommend that the fluid in the transmission be changed each 25,000–30,000 miles. The fluid is exchanged for new fluid and in some applications a filter is replaced as well. It is important to know what the steps are and why they are performed for you to be qualified to explain it to the customer. Some maintenance may also be needed to fulfill requirements of warranty, both standard and extended.

Task B.5.2 Identify related maintenance and reset procedures.

An example of a related item might be replacement of a water pump, if the timing belt drives it, during the timing belt replacement procedure. Since the water pump is accessible during the procedure and often fails right afterward if not addressed, it may be wise to recommend it at the same time. This would save your customer money and later failure that might be blamed on the timing belt replacement. One other example might be the replacement of the fuel filter when an in-tank electric fuel pump is being replaced. Because fuel pumps often shed metal and field coil winding when they are on their way out, the fuel filter can be restricted and should be replaced at the same time. To recommend and sell related items benefits your customer, as well as provides your business with the additional sale.

Task B.5.3 Locate and interpret maintenance schedule information.

Maintenance schedules are printed in the vehicle owner's manuals as well as in manufacturer's books and information systems. Interpreting the schedule requires knowing the mileage and type of use the vehicle experiences as well as knowing what procedures are required to maintain the systems. If the cooling system must be serviced at 60,000 miles, what does that service involve? The manufacturer may spell it out for you or leave it to standard procedures. In this case, we can say that our cooling system service would include replacing the coolant and verifying its protection, inspecting the condition of all hoses and the radiator, and verifying the correct operation of the thermostat. Some vehicles may require other things based on the experience of the vehicle's history. Tracking history and reviewing it upon service visits helps avoid mistakes in service suggestions.

Task B.6 Warranty, Service Contracts, Service Bulletins, and Campaign/Recalls (3 Questions)

Task B.6.1 Demonstrate knowledge of warranty policies and procedures/parameters.

To test your understanding of this task on an even playing field, the question will have to be posed as a scenario. All the facts will be given and you will have to make a decision based on the policy outlined in the question. The statement may be something like: A customer returns their vehicle for warranty on work that was performed 16 months ago. If the shop has a 12-month/12,000-mile warranty, what is the vehicle owner's responsibility? This question will be posed in the sample questions area so that you can see if you understand it.

Task B.6.2 Locate and use reference information for warranties, service contracts, service bulletins, and campaign recalls.

Depending on where you work, you may or may not work with service contracts. These are programs provided by various suppliers that extend the standard warranty. The process behind most of these programs is similar to working with an insurance company. The shop or the vehicle owner initiates a claim for needed repairs; an estimate is provided; the company providing the contract approves the part of the repair order that is covered; and the shop performs the work.

Warranty work and manufacturer campaigns/recalls are different. Warranty work may be paid for by the manufacturer in the case of new car warranty or may be the responsibility of the shop on a repair performed at their facility. Each business will have its own procedures for warranty. You should be up-to-speed on your warranty policies and know where to access written policies for your customer.

When a manufacturer finds an engineering problem or failure of a specific component in some kind of pattern, they may issue a campaign or a recall to repair or replace that component. Generally these have an associated technical service bulletin to help the shop performing the work understand the issue and the procedures associated with it. Because campaigns and recalls are generally paid for by the manufacturer, they are nearly always performed by the manufacturer's dealer.

There are several sources for the information on campaigns and recalls. They include the manufacturer's information system, aftermarket information systems, trade magazines, the National Highway Traffic Safety Administration web site, and customer letters sent by the manufacturer.

Task B.6.3 Demonstrate knowledge of service contracts, technical service bulletins, and campaign/recall procedures.

The natural extension of the last task is to explain to our customers what we are doing to their vehicles. You will notice the addition of technical service bulletins because they almost always have a service procedure attached to them. Include a written signature authorization as legal proof of explanation for a verbal authorization.

Task B.6.4 Verify applicability of warranty, service contracts and campaign recalls.

Because these types of repairs are often VIN specific, it is important to make sure that the bulletin or campaign applies to the vehicle on which you are working. Secondly, make sure warranties and service contracts are in force for the vehicle before performing repair using them. This may require a call to the manufacturer or the contract provider.

Task B.7 Vehicle Identification (2 Questions)

Task B.7.1 Locate and utilize vehicle ID number (VIN).

At the beginning of the 1980s, vehicles had become very diverse; even within the same bodyline there could be many variations. It had also become apparent that with vehicles sold in the United States coming from so many different origins, there was a greater risk that two completely different vehicles might end up with the same serial number. It was at this time that the 17-digit VIN system we use today was adopted. The VIN became a code to determine the make, model, body, origin, year of production, engine size, and production sequence number. Using this code ensured that each vehicle would have a unique identification sequence. There are locations within the VIN that are specifically used to identify certain information. Others are left to the manufacturer to use as they wish. Look at some of the required VIN digits. The first position is used to denote the origin and manufacturer of the vehicle. Letters and numbers have been used. The eighth position is the engine for all domestic manufacturers and the tenth digit is the year of production for the vehicle. The last six or seven digits are the serial number. It is very important that the VIN is recorded correctly and kept on file.

Task B.7.2 Locate production date.

Familiarize yourself with the location of the production date on the vehicle lines you work on in your shop. You will find that most manufacturers have a data plate or sticker inside the driver side door jamb to give you information about the vehicle. Some vehicles have the production date on a plate under the hood, usually on the fire wall. The various locations make it unlikely that you will see a question asking where a production date is. Correct part ordering requires this forgotten information.

Task B.7.3 Locate and utilize component identification data.

Many vehicle components have tags or markings that identify them when it comes time to replace them. These may be part numbers or engineering numbers cast into a part or a sticker on the component. These are often invaluable when ordering parts. A simple battery sticker can provide much information as to spec, size, group, and age.

Task B.7.4 Identify body styles.

There are a number of ways to identify a vehicle's body style. For models with limited number of variations available, you can simply identify it as a two-door, four-door sedan, station wagon, etc. Some manufacturers have more than 20 variations in a single body design. This is where it becomes important to use the VIN and the badges on the body to identify the vehicle. Modern software makes suggestions and provides drop-down displays to properly enter a vehicle.

Task B.7.5 Locate paint and trim code(s).

As with the production date, the paint and trim codes are often found on the door jamb, or under the hood on a plate. Some may even be located on the glove box door or the spare tire cover. You will need to be able to find the location of this information. Express care and insist that your techs perform steps to preserve, not destroy, any of the above identifications. These become a critical part of the repair process.

C. Sales Skills (13 Questions)

Task C.1 Provide and explain estimates.

After we have assembled a complete and accurate estimate for work needed on a customer's vehicle. it is time to present it to them by phone or in person. Depending on your presentation at this time, you may or may not have an easy job of checking out the customer when they come to pick up the vehicle. The best time to reach a full understanding with a customer is before the work is done. This way if you find that you cannot help the customer agree to the value of performing a certain service, you can remove it from your estimate. It is an excellent approach to ask the customer how much information they would like you to provide about the work estimated. While some customers may want considerable detail and explanation, you may actually lose a sale with some customers who do not want the information in great detail. Be willing to adapt your presentation to the customer's personality type. If the customer asks a question for which you do not have an answer, offer to get the information and get back to them. Too many service consultants provide incorrect information because they are in a hurry. This will almost always start a web of confusion that will end in the customer's distrust of the service consultant and ultimately the repair facility. It is of critical importance that the items on the repair order include all costs associated with them. If you offer an estimate that does not include shop supplies, environmental charges, tax, etc., the customer will perceive this as dishonesty. Selling the estimate to the customer is as much about selling yourself as it is selling the work. People buy service from people not from businesses. Where possible, attempt to stagger work drop-off and delivery times. This helps to ensure that you will have the time to explain all that is necessary.

Task C.2 Identify and prioritize vehicle needs.

When preparing an estimate, it is important to keep in mind which services are absolutely critical now, which are discretionary, and which are more cost-effective when grouped with other services. When you write up your estimate knowing this, you are in a better position to help the customer prioritize the work they want done. Many customers do not have the discretionary funds to perform all of the service at one time.

The service consultant has the opportunity to become the hero by helping to point out savings with bundled services or items that can be rescheduled for a later date. Sometimes the business has extended financing options that can make it possible to do all of the work at one time. Remember, control the conversation but give the customer options. You do this every day and have more experience with financial and repair strategies. If you offer good, sound options you are more likely to close a sale. Your focus should be on the first priority; that is, what the vehicle was brought in for.

Task C.3 Address customer concerns.

This can mean several things. When we offer complete explanations and solutions to the customer's concerns on the repair order, we are addressing their concerns. We are also addressing their concerns when they have questions about the reason and expectations from a given repair. Addressing a customer's concerns is never a negative; it is an opportunity to make them feel at ease with you, the business, and the work to be performed. Treat customers as you would like to be treated. Address problems the same way you treat new sales. This adds credibility.

Task C.4 Communicate the value of selling related and additional services.

Just as you had to learn how the many vehicle systems interrelate to one another, you must help the customer to understand why you might, for example, flush the power steering system when you are replacing a power steering pump. You must help them to see the benefits of services that will really complete a job. Most people dislike being without their car, which causes them to try to only leave it off when absolutely necessary. It is important for the repair business to keep this in mind and offer and anticipate needed services while a vehicle is in for other repairs. This will keep the customer on the road, which is a win for both the customer and the shop. It also helps avoid failed parts or additional repairs due to incompleteness of the original sale and repair.

Task C.5 Explain product/service features and benefits.

Features are the list of items that define a product while benefits are the tangible or perceived items the customer takes along with the product. In the communications section, we talked about features and benefits. Let us look at our spark plug scenario again.

A feature of a spark plug might be that it has a long-lasting design. The benefit is that the customer will not have to replace it as often. The plug's life is of no consequence unless there is a perceived or tangible value to that long life. That value comes from the fact that the expense of replacing spark plugs as well as the more frequent visit to the shop is lessened by the design. Therein lies the benefit. When you sell products or services to customers, be sure that you include the benefits in your sales pitch. Another way to think of it is that the benefit is what the product does for you or how it makes you feel. Simple, laminated, colorful visual aids add value to your explanation and help create understanding in customers.

Task C.6 Overcome objections.

When a customer objects to a needed repair there may be several reasons. It is up to the service consultant to find out why. Usually the best approach is to ask them why they do not want the service performed. It is possible that they have had it done just recently, which is something you should know. Nothing undermines a service consultant's credibility more than recommending a repair that has just been done (or that the customer thinks has just been done). If you do not know what the customer thinks, you cannot address the objection. Another common reason for an objection is that the customer does not understand the operation or see the value in it. Here again, asking questions to find out what the concern is will help you to fill in the blanks for them. You still may not make the sale, but the fact that you made the effort to explain your position goes a long way toward customer trust. History tracking or asking some initial qualifying questions can help avoid service suggestion pitfalls.

Task C.7 Close sale.

There are many ways to take your own best efforts and lose a sale. One way service consultants lose sales is they feel that when a customer is quiet they must continue to sell. Chances are the customer is mentally checking their finances or weighing which bill they will have to be late paying in order to pay you. One thing service consultants need to learn to do is stop talking after they make their pitch. Another common mistake is not asking for the sale. The customer is going to separate themselves from part of their income.

Remember we said that people buy service from people. It can really help to simply say, "We would be happy to perform these repairs for you, if you would like." If the customer suggests they would like a second opinion, a good response might be, "Yes, I think it is always a good idea to get a second opinion. I think our diagnosis and pricing is fair and consistent with other repair facilities, but I would be happy to offer you a printout of our diagnosis to have another shop confirm it for you." Most customers are going to decide that if you are confident enough to allow another shop to confirm your diagnosis they might as well have the work done now as opposed to going through the cost and time to do another diagnosis. Trust is key to the relationship between the consultant and the customer. Offer options, don't pressure, and speak knowledgeably and in a professional manner.

D. Shop Operations (3 Questions)

Task D.1 Manage work flow.

The service consultant is in the unique position of knowing the customer's needs and the technician's work load. This puts him in the perfect position to manage the work flow in the shop. By keeping track of the amount of time necessary to complete a customer's vehicle along with parts availability, the service consultant can determine the best times to promise completion to the customer. Once these expectations are set, it is important to stay on top of the work flow to be sure that technicians are receiving parts as expected and are able to complete the work on time. This is an area where a little communication can make the service consultant's job much easier. Shop efficiency relates directly to a shop's profitability or inability to support shop expenses.

Task D.2 Use available shop management systems (computerized and manual).

When developing the task list for an ASE test, the panel tries to outline the minimum requirements an individual must possess to be considered a viable employee. This task asks you to use available shop management systems. These systems print out work orders that are easy to read, calculate without error, and spell correctly. With well over 100 manufacturers of computerized systems out there and over 85% of repair facilities using computer-generated repair orders, this task does indeed reflect a skill you must possess, but it does not avail itself of a way to test candidates from all different types of service facilities. The most likely use of this task within the test would be in an overlap with another task.

Task D.3 Identify labor operations.

As promised, this task overlaps other areas where you identify labor operations to customers and in your estimating procedure. Tasks A.1.10, A.2.2 and A.2.6 will cover this task.

Task D.4 Demonstrate knowledge of sublet procedures.

Many shops sublet operations that are not performed in their facilities. Common examples would be driveline repair/balancing, transmission overhaul, body and paint work, or radiator repair. Even if your shop does perform some of these services, you should be aware that not all do and be able to answer questions about them. Example questions will be provided in the sample tests. Knowledge builds trust through effective repairs.

Task D.5 Maintain customer appointment log.

Most of the repair shops in this country schedule appointments for their customers. If your facility does not, you will still be expected to know how. When a customer calls for an appointment, the service consultant must determine what their needs are, about how much time the service consultant expects it to take to repair the vehicle, and the availability of a technician(s) with adequate skills to complete the repair. An appointment log offers an organized method for scheduling, and also reminds the consultant of incoming work.

Task D.6 Address repeat repairs/comebacks.

Having to repair the same problem more than once can kill shop productivity and morale. When these problems present themselves, the service consultant is often called on to get to the bottom of the problem and come up with an appropriate response. Is the technician missing the root of the problem for some reason? Is a part failing repeatedly? Does the customer keep bringing the vehicle back with a problem that you think is resolved? Here is where the service consultant's question-asking expertise comes in. By interviewing the customer and the technician, the service consultant can determine the best course of action and explain it to all involved. These types of problems are usually because a key piece of information is missing–a symptom the customer left out or a technical service bulletin that outlines a procedure to alleviate the repeat problem. The service consultant must work with the technician and the customer to find the problem's resolution. As with all things, doing a little research the first time the problem presents itself can save a customer and a lot of the shop's resources later. Some modern day problems do require a second or follow-up visit, so educating your customer on intermediate problems may result in a second chargeable invoice being sold with an incident. Always treat comeback customers with the same courtesy you provided when they first visited.

5 Sample Test for Practice

Sample Test

Please note the letter and number in parentheses following each question. They match the task in Section 4 that discusses the relevant subject matter. You may want to refer to the overview using the cross-referencing key to help with questions posing problems for you.

1. In most general repair shops, which of these is LEAST Likely to be a sublet operation?
   A. Windshield replacement
   B. Spark plug replacement
   C. Automatic transmission overhaul
   D. Drive shaft balancing (D.4)

2. A customer recites a list of symptoms to the service consultant. What should the service consultant do next?
   A. Write down exactly what the customer says.
   B. Use his/her experience to estimate repairs.
   C. Offer suggestions about what the problem might be.
   D. Ask open-ended questions to determine customer needs. (A.1.3)

3. A service consultant has prepared an estimate from a technician's diagnosis. Before providing the customer with the estimate, which of these should the service consultant perform first?
   A. Agree on a completion time with the technician.
   B. Verify availability of necessary parts.
   C. A thorough test drive.
   D. Identify additional maintenance needs. (A.2.3)

4. Service Consultant A says that when greeting a customer, the service consultant should offer their name and a handshake. Service Consultant B says that when greeting a customer, the service consultant should make eye contact and smile when welcoming them. Who is right?
   A. A only
   B. B only
   C. Both A and B
   D. Neither A nor B (A.1.6)

5. A fellow service consultant is upset with one of the shop's technicians. Which of these should the service consultant do?
   A. Encourage his fellow consultant to talk with the technician.
   B. Offer to work with that technician until the situation blows over.
   C. Speak to the technician himself.
   D. Alert the service manager immediately. (A.2.9)

6. A 30,000-mile maintenance procedure is being performed. Which of these is a benefit of performing the service?
   A. The cooling system gets flushed.
   B. The transmission fluid gets changed.
   C. The vehicle will continue to deliver dependable service.
   D. The completed checklist is given to the customer. (C.5)

7. A Service Consultant finds a need to provide alternate transportation to an under 21-year-old customer. Which of these would be the best option to offer this customer?
   A. Offer them a reduced rate rental car.
   B. Lend them the boss's vehicle.
   C. Offer directions to the nearest bus stop.
   D. Offer to provide a ride. (A.1.7)

8. Which of these is the primary function of the shock absorbers/struts in the suspension?
   A. Support the weight of the body for the suspension
   B. Control front to rear ride height trim
   C. Compensate for weight changes
   D. Dampen and control suspension movement (B.3.2)

9. A vehicle is being serviced for the first time and has had numerous repairs at both dealers and independent shops. Service Consultant A suggests telling the customer that the shop's technicians are much more competent than the other shops they have used in the past. Service Consultant B asks the customer to bring their repair records along to help the shop get an idea of the vehicle's history. Who is right?
   A. A only
   B. B only
   C. Both A and B
   D. Neither A nor B (A.1.9)

10. When a customer objects to the cost of a given repair, which of these is the best response by the service consultant?
    A. Offer the customer a discount on the repair.
    B. Reschedule for a later time.
    C. Explain the reasons for and benefits of the repair.
    D. Remove it from the repair order immediately. (C.6)

11. A vehicle in the shop for an oil change shows approximately 59,000 miles on the odometer. What should the service consultant do?
    A. Suggest an appointment for 60,000-mile maintenance.
    B. Offer the customer a discount to perform a 60,000-mile maintenance today.
    C. Advise that the 60,000-mile maintenance is covered under manufacturer's warranty.
    D. Provide a ballpark estimate for a 60,000-mile maintenance. (A.1.10)

12. Service Consultant A says that vehicles that do not have frames around the windows are known as hard tops. Service Consultant B says that a hatchback is a vehicle with a rear trunk/window combination that lifts up. Who is right?
    A. A only
    B. B only
    C. Both A and B
    D. Neither A nor B (B.7.4)

13. A customer has just given approval for repair of their vehicle. Service Consultant A says the technician should be provided with the approved work order. Service Consultant B says documentation of the customer's approval should be on the work order. Who is right?
    A. A only
    B. B only
    C. Both A and B
    D. Neither A nor B (A.1.12 and D.1)

14. A technician turns in a repair order that recommends replacement of the CV boot with no further description. Which of these should the service consultant do next?
    A. Estimate replacement of complete axle.
    B. Verify parts availability.
    C. Determine the reason for the repair.
    D. Check vehicle repair history. (A.2.6)

15. Each of these represents an example of customer information that might be included on a repair order **EXCEPT:**
    A. e-mail address.
    B. preferred method of payment.
    C. cell phone number.
    D. service consultant name. (A.1.4)

16. Service Consultant A says that adding a description of the work performed adds value to the repair. Service Consultant B says that the VIN may be used to find part applications. Who is right?
    A. A only
    B. B only
    C. Both A and B
    D. Neither A nor B (A.2.7 and B.7.1)

17. A customer enters the service area while the service consultant is on the telephone with another customer. Which of these should the service consultant do?
    A. Finish the conversation with the telephone customer first.
    B. Place the telephone customer on hold to take care of the walk-in customer.
    C. Acknowledge the walk-in customer with a wave and finish with the telephone customer.
    D. Ask the telephone customer if she may call them back. (A.1.1)

18. A customer calls for a fellow service consultant who is already working with a customer. Which of these should the service consultant do?
    A. Take the customer's name and number and promise a call back.
    B. Attempt to help the customer.
    C. Place the customer on hold until the consultant is available.
    D. Transfer the call to the owner/service manager. (A.1.4 and A.2.9)

19. When booking an oil change appointment for a good customer, the service consultant finds that his drivability technician is the only one with openings on the day the customer wants. Which of these is the best solution to this problem?
    A. Book the job for the drivability technician.
    B. Add it to the lube tech's schedule and tell the customer you will work them in.
    C. Offer the closest time that does not have a conflict.
    D. Move the appointment of a new customer. (A.2.8 and D.5)

20. All of these are components of the charging system **EXCEPT:**
    A. the starter.
    B. the battery.
    C. the voltage regulator.
    D. the alternator. (B.1.3)

21. Service Consultant A says that many automatic transmissions do not have a filter that is serviced during normal service. Service Consultant B says that there are many types of automatic transmission fluids that must be identified before service. Who is right?
    A. A only
    B. B only
    C. Both A and B
    D. Neither A nor B (B.2.1)

22. An automatic transmission is being replaced. Service Consultant A says that the transmission cooler should be flushed prior to connecting to the transmission. Service Consultant B says that the engine oil must be changed to guarantee good transmission life. Who is right?
    A. A only
    B. B only
    C. Both A and B
    D. Neither A nor B (B.2.3)

23. When a vehicle is found to need maintenance work that was not requested by the customer, a service consultant should recommend it because:
    A. it provides additional income for the repair facility.
    B. it is the responsibility of the repair facility to advise them of their vehicle needs.
    C. customers must have this work done to maintain their warranty.
    D. it will keep the vehicle in good working order. (A.1.8 and C.4)

24. Which of these might be performed along with cooling system flush as a related or overlapping item?
    A. Replacement of thermostat
    B. Cooling fan replacement
    C. Battery service
    D. Cabin filter replacement (B.5.2)

25. A customer calls and states that their vehicle has a problem that has had several repair attempts. Which of these should the service consultant do first?
    A. Determine if the dealership/shop has ever worked on the vehicle.
    B. Offer to take the vehicle in immediately.
    C. Ask the customer to provide previous work orders.
    D. Explain that sometimes a problem can take several attempts to resolve. (A.1.13)

26. Which of these is a component of the antilock brake system (ABS)?
    A. Wheel cylinder
    B. Torque converter
    C. Wheel speed sensor
    D. Throttle position sensor (B.3.1)

27. A customer has come to pick-up their vehicle when the service department is very busy. Which of these is the best way to handle the situation?
    A. Direct the customer to the cashier/cash them out.
    B. Advise the customer that you are very busy.
    C. Review the work performed and the invoice with the customer.
    D. Ask them to come back when it is quieter. (A.1.16)

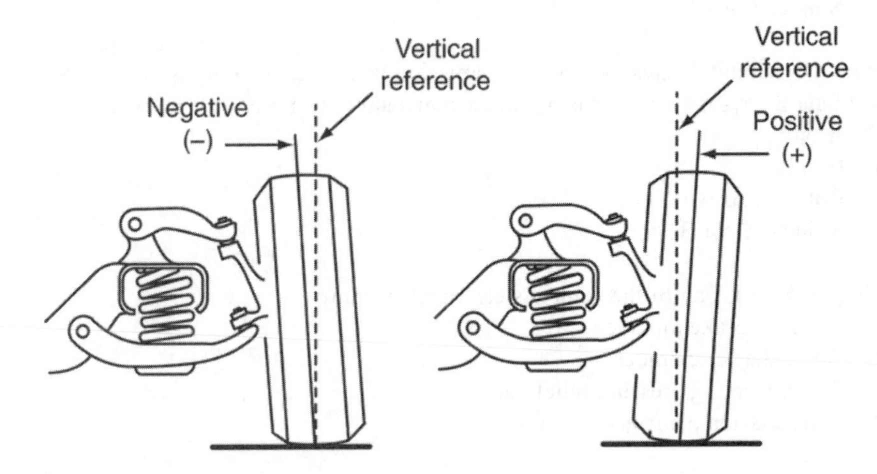

28. In the figure, the wheel alignment angle indicated is:
    A. thrust angle.
    B. camber.
    C. caster.
    D. toe out. (B.3.3)

29. After inspecting a vehicle, the technician recommends the following: replacement of a damaged driver side seat belt, cooling system flush, replacement of brake pads that have 3⁄32 inch remaining, and oil change that is 1,500 miles overdue. Which of these represents the best way to prioritize this list to the customer?
    A. Brake pads, oil change, seat belt, cooling system service
    B. Seat belt, oil change, brake pads, cooling system service
    C. Seat belt, brake pads, oil change, cooling system service
    D. Oil change, cooling system service, brake pads, seat belt (C.2)

30. Service Consultant A says that the air conditioning condenser converts refrigerant from gas to liquid. Service Consultant B says that air conditioner compressors create both vacuum and pressure. Who is right?
    A. A only
    B. B only
    C. Both A and B
    D. Neither A nor B (B.4.1)

31. Which of these is a component of the ignition system?
    A. Coil pack
    B. Fuel filter
    C. Thermostat
    D. Throttle position sensor (B.1.1)

32. Service Consultant A says that providing an estimate is required by law in most states. Service Consultant B says that explaining the details of the estimate helps to add value to the services the customer is buying from the shop. Who is right?
    A. A only
    B. B only
    C. Both A and B
    D. Neither A nor B (C.1)

33. Service Consultant A says the CVT transmission is an automatic transmission. Service Consultant B says the CVT transmission is a manual transmission. Who is right?
    A. A only
    B. B only
    C. Both A and B
    D. Neither A nor B (B.2)

34. Which of these is part of the vehicle's electrical system?
    A. Antilock brake module
    B. Brake master cylinder
    C. Power steering pressure relief valve
    D. Transmission planetary (B.4.2)

35. When starting the vehicle, which of these does the starter engage?
    A. The alternator
    B. The starter drive
    C. The starter solenoid
    D. The battery (B.1.2)

36. Service Consultant A says that providing a ballpark estimate is a useful tool to closing a sale. Service Consultant B says that being friendly and asking for an appointment is a good way to close a sale. Who is right?
    A. A only
    B. B only
    C. Both A and B
    D. Neither A nor B (C.7)

37. Which of these is a part of a cooling system service?
    A. Verifying pH, protection, and pressure testing the system for leaks
    B. Replacing thermostat
    C. Heater core replacement
    D. Adding R134A as needed (B.5.1)

38. Service Consultant A says that maintenance schedules are printed in the vehicle owner's manual. Service Consultant B says that maintenance schedules are selected based on the customer's use of the vehicle. Who is right?
    A. A only
    B. B only
    C. Both A and B
    D. Neither A nor B (B.5.3)

39. Service Consultant A says that telling the customer when their vehicle will be ready at the time they drop off the vehicle creates expectations. Service Consultant B says that accurate completion times can only be determined after vehicle inspection. Who is right?
    A. A only
    B. B only
    C. Both A and B
    D. Neither A nor B (A.1.11)

40. Each of these describes a purpose of a technical service bulletin **EXCEPT:**
    A. a document mailed to customers to let them know about a problem with their vehicle.
    B. a document that explains a redesign of a component.
    C. a document that revises a shop manual procedure.
    D. a document that describes a pattern failure in a vehicle or group of vehicles. (B.6.3)

41. Which of these is NOT needed to determine applicability of a vehicle's service contract?
    A. Mileage
    B. Vehicle identification number (VIN)
    C. In-service date
    D. Production date (B.6.4)

42. Which of these is to have the greatest impact on a customer's decision to do business with you?
    A. Extended business hours
    B. The service consultant's appearance
    C. The level of trust they feel
    D. Discount pricing (A.1.14)

43. Which of these is a common location for the production date?
    A. Stamped on the valve cover
    B. Inside the driver door pillar
    C. Sticker on the radiator support
    D. Inside the gas door (B.7.2)

44. Which of these is a component of the SRS or air bag system?
    A. Throttle sensor
    B. PCM
    C. Wheel speed sensor
    D. Clock spring/spiral cable (B.4.3)

45. During quiet times when customer and phone demands are low, a service consultant takes time to visit with each technician working on customer's vehicles for which the consultant is responsible. Service Consultant A says that this helps keep track of progress within the shop. Service Consultant B says that this helps to identify jobs that may not be completed as promised. Who is right?
    A. A only
    B. B only
    C. Both A and B
    D. Neither A nor B (D.1)

46. A customer receives a letter from the manufacturer for which of these actions?
    A. A technical service bulletin
    B. A vehicle campaign
    C. End of vehicle warranty
    D. A vehicle recall (B.6.1)

47. Service Consultant A says that when writing up a comeback/warranty ticket it is necessary to review previous repair orders with the customer. Service Consultant B says that when writing up a comeback/warranty ticket it is necessary to ask the customer to restate the symptoms they are experiencing. Who is right?
    A. A only
    B. B only
    C. Both A and B
    D. Neither A nor B (A.1.2, A.1.9 and D.6)

6 Additional Test Questions for Practice

Additional Test Questions

Please note the letter and number in parentheses following each question. They match the task in Section 4 that discusses the relevant subject matter. You may want to refer to the overview using the cross-referencing key to help with questions posing problems for you.

1. Service Consultant A says that an example of a feature of an oil change is the brand of oil used. Service Consultant B says that an example of a benefit of an oil change is longer engine life. Who is right?
   A. A only
   B. B only
   C. Both A and B
   D. Neither A nor B (C.5)

2. The catalytic converter is always located:
   A. in the tail pipe.
   B. behind the muffler.
   C. near the air pump.
   D. behind the primary oxygen sensor. (B.1.1)

3. Supplemental restraint system (SRS) is another name for:
   A. seat belts.
   B. air bags.
   C. upper hydraulic motor mount.
   D. child safety seat anchors. (B.4.3)

4. Which of these is NOT an emission control device?
   A. PCV valve
   B. Fuel injector
   C. Air pump
   D. Fuel cap (B.4.1)

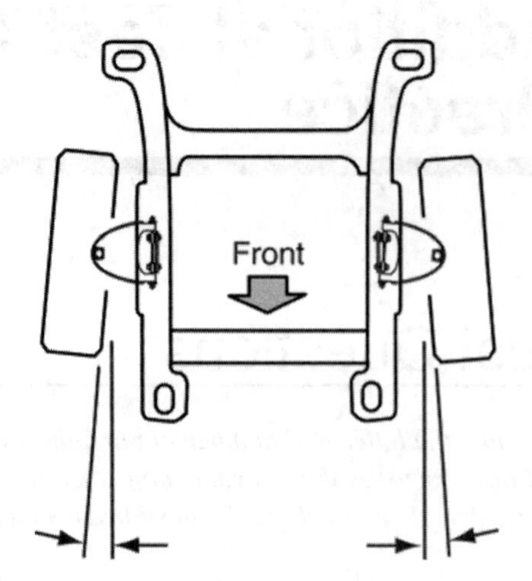

5. In the figure, what is the wheel alignment angle indicated?
   A. Thrust angle
   B. Camber
   C. Caster
   D. Toe (B.3.2)

6. When a customer is picking up their vehicle, Service Consultant A says that it is important to take the time to explain the work performed in as much detail as the customer requires. Service Consultant B says that if the customer asks questions it indicates they do not trust the shop/dealership. Who is right?
   A. A only
   B. B only
   C. Both A and B
   D. Neither A nor B (A.1.16)

7. Service Consultant A says a customer that only drives 500 miles a month should have the vehicle serviced based on the miles driven. Service Consultant B says that a customer that drives 5,000 miles a month should service the vehicle based on the miles driven. Who is correct?
   A. A only
   B. B only
   C. Both A and B
   D. Neither A nor B (B.5.3)

8. A service consultant has just completed compiling and writing up a customer's concerns. Which of these should the consultant do next?
   A. Dispatch the work order to the technician.
   B. Arrange for a ride home for the customer.
   C. Offer an estimate for the repairs needed.
   D. Confirm the accuracy of the information with the customer. (A.1.2 and A.1.5)

9. Service Consultant A suggests that offering a customer a ride home or to work represents alternative transportation. Service Consultant B suggests that driving the customer to the bus stop is providing alternate transportation. Who is right?
   A. A only
   B. B only
   C. Both A and B
   D. Neither A nor B (A.1.7)

10. A technician recommends alignment due to excessive negative camber. Which of these best explains this condition?
    A. The front wheels are both pointing outward.
    B. The wheel is tilted out at the top.
    C. The wheel is tilted in at the top.
    D. The spindle is tilted back at the top. (B.3.2)

11. Which of these is NOT present when a vehicle has a DIS ignition system?
    A. Fuel filter
    B. Distributor
    C. Cam sensor
    D. Ignition coil (B.1.3)

12. Spark plugs fire on both power and exhaust stroke in which of these ignition systems?
    A. Distributorless
    B. Point-type distributor
    C. Electronic distributor
    D. Diesel (B.4.1)

13. Service Consultant A says that a vehicle with a non-interference engine may bend valves if the timing belt breaks. Service Consultant B says that the timing belt drives the camshaft(s). Who is right?
    A. A only
    B. B only
    C. Both A and B
    D. Neither A nor B (B.1.1 and B.1.3)

14. The technician notes on the repair order that the fuel filter appears to be original on a vehicle with almost 60,000 miles on it. The item calls for replacement at 30,000 miles. Which of these should the service consultant do?
    A. Estimate a maintenance tune-up including the fuel filter.
    B. Ask the customer when and if it was replaced.
    C. Tell the customer that it has not been replaced in 60,000 miles.
    D. Leave the item for the 60,000-mile maintenance. (A.1.10)

15. Each of these might use a filter **EXCEPT:**
    A. automatic transmission.
    B. brake master cylinder.
    C. air-conditioning system.
    D. vehicle cabin. (B.2.3)

16. The connecting rod connects to the crankshaft at one end and the:
    A. flywheel.
    B. rear main seal.
    C. piston.
    D. cylinder head. (B.1.1)

17. A customer comes in with multiple complaints/concerns with limited funds. The service consultant should:
    A. Write up the complaints until all the customers' money is used.
    B. Pick out the easiest jobs that cost the customer the most money first.
    C. Recommend the customer save there money and come back when they have enough for all the repairs.
    D. Prioritize the repairs with the important ones first and let the customer choose how much they can afford to do now and schedule the additional repairs. (C.2)

18. A customer wants to have 4 new tires installed and the technician finds a bent rim that will cause vibration and tire wear problems and you find the rim is no longer being made. What should the service consultant recommend to the customer?
    A. I will try and find a replacement rim at a salvage yard.
    B. The rim is no longer available to match your old rims and recommend you choose 4 new rims that match.
    C. I say you should go ahead and install the new tires and see how it feels.
    D. The rim could fly apart and cause an accident so you should replace all the rims. (B.3)

19. An upset customer comes in when the service department is very busy. Which of these is the best way to handle the situation?
    A. Listen to them explain the whole problem.
    B. Ask them to come back after they have cooled down.
    C. Point out the areas in which they are wrong.
    D. Offer to discount the repair. (C.6)

20. All of these are components of an automatic transmission **EXCEPT:**
    A. planetary.
    B. pressure plate.
    C. torque converter.
    D. valve body. (B.2.3)

21. During a cooling system service, Service Consultant A says that air-conditioning system output temperature should be measured before and after the service. Service Consultant B says the type of antifreeze used in the system may be identified by color. Who is right?
    A. A only
    B. B only
    C. Both A and B
    D. Neither A nor B (B.1.2)

22. The starter turns the engine by engaging which of these components?
    A. Transmission
    B. Flywheel ring gear
    C. Crank shaft
    D. Battery (B.1.3)

23. A customer complains that his new car that has an CVT automatic transmission won't shift. What should the service consultant ask or say next?
    A. When would you like to drop it off to have it fixed?
    B. You should call the manufacture and ask if they know why.
    C. Have you read your owners manual for the proper operation of your CVT transmission?
    D. Do not worry about it and call me when you have a real problem. (B.2)

24. The CV axle connects the differential or transaxle to which of these?
    A. Wheel hub
    B. Tires
    C. Engine
    D. Flywheel (B.2.3)

25. When writing up a customer's work order, which of these is the first thing to ask for?
    A. The vehicle VIN
    B. The customer's name
    C. The prime item
    D. The vehicle license number (A.1.2, A.1.4 and A.1.6)

26. A potential customer calls very concerned about an estimate received by another shop. Which of these should the service consultant do?
    A. Suggest that the other shop is probably too high and make an appointment.
    B. Look the job up and offer an estimate.
    C. Offer a discount if they bring the vehicle into your shop.
    D. Show empathy for them and offer an appointment for a second opinion. (A.1.1 and C.3)

27. The master cylinder is part of which system?
    A. Brakes
    B. Engine
    C. Drivetrain
    D. Steering (B.3.1)

28. A customer calls with a shopping list of problems with their vehicle. How does the service consultant put this information in a form that will help the technician find the customer's problem?
    A. Write down everything the customer says in the order they say it.
    B. Ask open-ended questions regarding each item to determine the problem.
    C. Ask the customer to boil the problem down to the area of the car they are talking about.
    D. Verify that each item on the repair order is a symptom or a maintenance request. (A.1.3 and A.2.1)

7 Appendices

Answers to the Test Questions for the Sample Test Section 5

1. B	13. C	25. A	37. A
2. D	14. C	26. C	38. C
3. B	15. D	27. C	39. C
4. C	16. C	28. B	40. A
5. A	17. C	29. B	41. D
6. C	18. A	30. C	42. C
7. D	19. C	31. A	43. B
8. D	20. A	32. C	44. D
9. B	21. C	33. A	45. C
10. C	22. A	34. A	46. D
11. A	23. B	35. B	47. C
12. C	24. A	36. D	

Explanations to the Answers for the Sample Test Section 5

Question #1
Answer A is wrong. Windshield replacement is usually sublet. It requires special tools and training.
Answer B is correct. If you are not doing spark plugs in a general repair shop, you are probably working on diesels only. Remember that ASE tests are designed to fit people working in parallel environments, so look for the use of "most," "generally best," or "most-likely" to indicate that you should have an open mind.
Answer C is wrong. Many shops do removal and installations only. The majority sublet the overhaul.
Answer D is wrong. It requires specialized equipment, not routine repair.

Question #2
Answer A is wrong. Customers are not technicians; ask questions.
Answer B is wrong. No need for an estimate, instead repair information should be given.
Answer C is wrong. Remain a consultant, and don't try to be the technician. You must allow time to diagnose.
Answer D is correct. By asking the customer questions that help them to flesh out their description of the problem, the service consultant can save both the customer and the technician time and money diagnosing the problem.

Question #3
Answer A is wrong. This would be done after you made sure that the parts would be available to complete the repairs.
Answer B is correct. The other items should be done at various times during the repair process, but before you decide on a completion time to tell the customer, check to see when you can have the parts in your hands. How much and how long are the most asked questions by customers.
Answer C is wrong. This might be done prior to estimating by either the service consultant or the tech.
Answer D is wrong. Though true, not first in complete sale approach.

Question #4
Answer A is wrong. Service Consultant B is also correct.
Answer B is wrong. Service Consultant A is also correct.
Answer C is correct. Both Service Consultants are correct. Courtesy and confidence will encourage your customer to reciprocate and make them feel at ease with you and your facility.
Answer D is wrong. Both Service Consultants are correct in their initial approach and contact.

Question #5
Answer A is correct. Within the shop, personality conflicts can happen. It is important to handle them quickly and in person. Service consultants must be the leaders in these situations because they tend to have more inter-personal skills training than technicians. If the problem cannot be resolved independently, then it would be wise to involve a manager or owner.
Answer B is wrong. This may further offend the technician and solve no problems.
Answer C is wrong. This is not this consultant's concern. This could add problems to the conflict.
Answer D is wrong. This would not be the first remedy. Owners and upper managers only want issues brought to them that have been refined and that you have already attempted to solve. This is the last resort, not the first.

Question #6
Answer A is wrong. This is a feature of the service not a benefit.
Answer B is wrong. This is a feature of the service not a benefit.
Answer C is correct. The other answers were examples of features. In themselves they do not have much value unless the customer understands the long range benefit the services provide. Dependability is a benefit and something you receive from the service.
Answer D is wrong. Both A and B are features of the performed service.

Question #7
Answer A is wrong. Many 21-year-olds will not qualify for a rental car.
Answer B is wrong. This could be a risk to the company and your employment.
Answer C is wrong. This is not the most courtesy offer.
Answer D is correct. With younger customers, it is important to remember that the facility's insurance may not cover them. This was the best and safest alternate transportation to offer these customers.

Question #8
Answer A is wrong. Springs support weight. They are not absorbers of any kind.
Answer B is wrong. Not the job of struts or shock absorbers.
Answer C is wrong. This would be done by the vehicle metal leaf support springs, air support springs or the coil springs.
Answer D is correct. We were looking for you to understand that A, B, and C are all the responsibility of the springs. The shocks/struts provide control to the suspension and dampen blows to the suspension or from rough roads.

Question #9
Answer A is wrong. The other technicians may be very competent. This may just be a difficult repair. The best technicians have difficult days.
Answer B is correct. Only Service Consultant B is correct. While it is possible that multiple shops have failed to repair a customer's vehicle, the wise service consultant does not rule out that the customer may be the problem or that the vehicle has a very difficult problem. When dealing with these, keep an open mind and don't be afraid to ask a lot of questions. The best technicians have difficult repairs.
Answer C is wrong. Only Service Consultant B is correct.
Answer D is wrong. Only Service Consultant B is correct.

Question #10
Answer A is wrong. Add value first, not discounts. Discounts become an expense to a shop.
Answer B is wrong. Work the objection out while the customer is present. If the customers leave they may never return.
Answer C is correct. The most common reason why customers decline needed work is that they do not understand the need for it or see the value in it. Your explanation is the best sales tool to overcome objections.
Answer D is wrong. The service consultant must maintain control of the repair and the sale. Add value.

Question #11
Answer A is correct. Getting the customer prepared and asking for the appointment is the best bet. The customer may ask you to do the work while it is there.
Answer B is wrong. Depending on the situation, you may be offering a discount to someone who would have the work done at regular price.
Answer C is wrong. This is not a warranty covered item (with very rare exceptions). If you are not a dealer you may be suggesting the work be done elsewhere.
Answer D is wrong. Guessing on prices can be a disaster. Customers always remember the lowest price in any range you give them.

Question #12
Answer A is wrong. Service Consultant B is also correct.
Answer B is wrong. Service Consultant A is also correct.
Answer C correct. Both Service Consultants are correct. There are many body styles. Since there can be many body styles in a single model of vehicle, it is a good idea to be able to identify them visually. Remember, if the door glass closes directly against a weather strip, the vehicle is a hard top, unless it is a convertible. Body styles can be mixed together, as in the 5-door hatchback that is a sedan and a hatchback.
Answer D is wrong. Both Service Consultants are correct.

Question #13
Answer A is wrong. Service Consultant B is also correct.
Answer B is wrong. Service Consultant A is also correct.
Answer C is correct. Both Service Consultants are correct. One area that may appear on the C1 test is documenting the customer's approval of the estimate. If it appears, it will be in this task. The method of documenting this approval includes the name of the person who gave consent, the time and date, the amount, and the method used to make contact, i.e., phone, in person, etc.
Answer D is wrong. Both Service Consultants are correct.

Question #14
Answer A is wrong. You must verify the need for the repair first.
Answer B is wrong. This is necessary only after the sale and authorization.
Answer C is correct. When a technician hands in a repair order that is incomplete, you must get the complete story so that you do not end up the center of misinformation that ends with you looking dishonest.
Answer D is wrong. Also not a bad idea but not the best answer of the choices you had.

Question #15
Answer A is wrong. An E-mail address is a customer contact method and is definitely customer information.
Answer B is wrong. Many shops ask for the customers preferred method of payment when they work with warranty repairs or aftermarket warranty companies.
Answer C is wrong. A cell or mobile phone is a customer contact method and is definitely customer information.
Answer D is correct. Indeed, you may put the service consultant's name on the repair order; it does not constitute customer information.

Question #16
Answer A is wrong. Service Consultant B is also correct.
Answer B is wrong. Service Consultant A is also correct.
Answer C is correct. Both Service Consultants are correct. These two are loosely talking about repair orders and the information on them. Customers may not read all of the description on the repair order but it can save you when that "ever since you ..." call comes in. It also adds value because you have described the process necessary to find and repair a problem. The VIN offers information for correct part ordering.
Answer D is wrong. Both Service Consultants are correct.

Question #17
Answer A is wrong. You should always acknowledge customers as they enter. You should finish your phone call as long as the customer in front of you gets a sign from you that you will get to them next.
Answer B is wrong. You should stay with the customer you were working with. The customer on the phone is just as important as the one in front of you.
Answer C is correct. Demonstrating courtesy would require some acknowledgment and this lets the customer know you see them and that they will be helped soon.
Answer D is wrong. Time for the phone caller is OK, just keep it brief and invite a visit. The phone customer should be taken care of in a timely manner.

Question #18
Answer A is correct. This is so important to customers. If you take a message, be sure you get it to the person it is intended for. This is also the best solution to avoid incomplete and rushed customer service.
Answer B is wrong. Your attempt to help might only cause confusion if there is already an open dialogue going on between your fellow consultant and the customer.
Answer C is wrong. This is what most people do, but it is not right. By placing the customer on hold, you are putting pressure on the other consultant to finish with the customer they are with. This effectively irritates two customers and may not leave time for a complete close to the consultant's current situation.
Answer D is wrong. Here again, the owner or service manager are not going to be any more help than you would be getting in the middle of the situation. They become the last resort to those in front of them in the employment chain.

Question #19
Answer A is wrong. Most service managers and shop owners frown on using the "high dollar" help to do lower margin work; other options are yet possible.
Answer B is wrong. This might work, after all they were only asking for an oil change and customers never add to the work order when they arrive, do they? Of course they do, and you don't want to lose the opportunity to provide for the additional work without turning your shop flow upside down. Not best answer.
Answer C is correct. The customer will not know your schedule so try to negotiate a mutually beneficial appointment time. This keeps the job profitable, and allows time for add on sales.
Answer D is wrong. Once you start a juggling act something will eventually fall. Leave your schedule alone and negotiate a time that works for both you and your customer.

Question #20
Answer A is correct. The starter is part of the starting/cranking system, not part of the charging system.
Answer B is wrong. The battery stores electricity created by the charging system and related components.
Answer C is wrong. The voltage regulator maintains system voltage and is a component of the charging system.
Answer D is wrong. The alternator is the primary component of the charging system.

Question #21
Answer A is wrong. Service Consultant B is also correct.
Answer B is wrong. Service Consultant A is also correct.
Answer C is correct. Both Service Consultants are correct. Consultant A is right that many automatic transmissions make use of a screen rather than a replaceable filter that is not serviced during fluid changes. Consultant B is also correct in that there are many different specifications of transmission fluids. Some may be compatible with many transmissions, and some are specific to a make or model of transmission.
Answer D is wrong. Both Service Consultants are correct.

Question #22
Answer A is correct. Only Service Consultant A is correct. The transmission cooler must be flushed or replaced to remove debris from the old transmission.
Answer B is wrong. Transmissions do not share oil with the engine.
Answer C is wrong. Only Service Consultant A is correct.
Answer D is wrong. Service Consultant A is correct.

Question #23
Answer A is wrong. Though true, the correct answer puts the customer first.
Answer B correct. While all of the answers have truth to them, the real answer is that we are the experts, and it is our responsibility to point out maintenance needs to our customers to help them avoid breakdowns.
Answer C is wrong. Not always true; could be a wrong statement.
Answer D is wrong. Is true, but is a result of suggested and sold additional maintenance by the service consultant.

Question #24
Answer A is correct. As we learned in Question 22 a thermostat is not part of a cooling system service, but it does represent a common related service that is performed along with a cooling system service. Related items are items that either are part of the same system being serviced, that work in conjunction with an item being serviced, or that offer some labor overlap if they are serviced at the same time.
Answer B is wrong. If needed, a cooling fan might be a related service when replacing a water pump or radiator because it might have to be removed to service them. There is no labor overlap in this repair.
Answer C is wrong. While both a cooling system flush and a battery service are maintenance items, they are not related and share no overlapping labor time.
Answer D is wrong. While both a cooling system flush and a battery service are maintenance items, they are not related and share no overlapping labor time.

Question #25

Answer A is correct. You will be much better equipped to handle the situation if you know whether your facility has ever worked on the vehicle. Service history is invaluable when problems occur.

Answer B is wrong. You may want to do this, but first find out if you have worked on the vehicle before you feel you must add strain to the shop schedule by bringing in an additional vehicle that obviously is not easy to fix. Additional time may be needed.

Answer C is wrong. This is correct, but not first; maybe some work was done elsewhere not in your history. You will want to ask the customer to provide you with the history of repairs for the problem and the vehicle to help your technician get a better idea of the problem.

Answer D is wrong. No customer wants to hear this. Whether you worked on the vehicle or not, try a more reassuring approach than this. You are the expert/professional.

Question #26

Answer A is wrong. The wheel cylinder is part of the brake system but not part of the antilock system.

Answer B is wrong. The torque converter is part of an automatic transmission.

Answer C is correct. Wheel speed sensor is a component of the ABS system and sends a signal to the ABS module to indicate wheel speed.

Answer D is wrong. The throttle position sensor is part of the engine control system.

Question #27

Answer A is wrong. Always find some time at the end of the sale to thank the customer.

Answer B is wrong. They care about their repair and want some of your time.

Answer C is correct. It may be your temptation to get paid and get to the next customer. If you want strong repeat sales, you have to take the time to check out your customers properly.

Answer D is wrong. A busy shop may never be quiet.

Question #28

Answer A is wrong. Thrust angle is the relationship between the overall toe direction of the rear wheels to the overall toe direction of the centerline of the chassis.

Answer B is correct. When you look at the wheel from the front, the amount the tire leans in or out at the top is called camber. When the tire leans out, it is called positive camber and, when it leans in at the top, it is called negative camber.

Answer C is wrong. Caster is an imaginary line drawn through the ball joints or from the top of the strut mount to the lower ball joint.

Answer D is wrong. Toe is the direction the front of the wheel points when going down the road. If you stand up and point your toes toward each other, that is what is known as toe in. If you point them away from each other, that is known as toe out.

Question #29

Answer A is wrong. The brake pads have no immediate need, and are actually third priority.

Answer B is correct. When a list of recommended items is provided by a technician, the service consultant must prioritize them based on need. Always suggest the safety items first, maintenance that is behind next, and discretionary items that can go for a while last.

Answer C is wrong. The oil change is 1500 miles overdue, while the brakes still have life.

Answer D is wrong. Safety items must be first and are the most immediate concern.

Question #30

Answer A is wrong. The word condense means just that (gas to liquid).

Answer B is wrong. As with an engine, a compressor uses a piston and aspirates air. Intake air (vacuum) and exhaust air pressure.

Answer C is correct. Both Service Consultants are correct. The Service Consultants are providing important basics of air conditioning theory. It is unlikely that a test question will appear on the test at this level, but, as a service consultant, you will find it much easier to explain expensive A/C system repairs to your customer if you understand system operation.

Answer D is wrong.

Question #31
Answer A is correct. The ignition coil takes a low voltage pulsing signal from the ignition module and amplifies is from about 9–12 volts to 5,000–40,000 volts and sends it to the spark plugs.
Answer B is wrong. The fuel filter is part of the fuel system.
Answer C is wrong. The thermostat is part of the cooling system.
Answer D is wrong. The throttle position sensor is part of the engine management system.

Question #32
Answer A is wrong. Service Consultant B is also correct.
Answer B is wrong. Service Consultant A is also correct.
Answer C is correct. Both Service Consultants are correct. Consultant A is correct. Most states have some sort of estimate law or motor vehicle repair act that requires shops to provide an estimate and receive approval before repairs are performed. Consultant B is also correct. Many times a sale is made on an item just because the consultant took the time to explain to the customer the value of having the repair done.
Answer D is wrong. Both Service Consultants are correct.

Question #33
Answer A is correct. Only Service Consultant A is correct. The CVT or continuously variable transmission is considered an automatic transmission.
Answer B is wrong. This kind of transmission is an automatic style transmission.
Answer C is wrong. Only Service Consultant A is correct.
Answer D is wrong. Only Service Consultant A is correct.

Question #34
Answer A is correct. This is the only electrical component in the list. It is helpful to understand that many systems make up the electrical system in addition to the battery, starter, and alternator.
Answer B is wrong. The master cylinder is a mechanical hydraulic component.
Answer C is wrong. The power steering pressure relief valve is a mechanical component.
Answer D is wrong. The planetary (gearset) is a mechanical device inside the transmission.

Question #35
Answer A is wrong. It is not related to the starter.
Answer B is correct. The starter drive is the only item here that the starter engages.
Answer C is wrong. The starter solenoid is certainly part of starting a vehicle but the starter does not engage it. The solenoid engages the starter.
Answer D is wrong. The battery stores electricity for starting only.

Question #36
Answer A is wrong. Ball-park estimates have no value and can be wrong.
Answer B is wrong. Close the sale first; then ask for the appointment.
Answer C is wrong. Friendly is OK, but all else is incorrect.
Answer D is correct. A ball-park estimate is a good way to get a case of 5 o'clock shock and lose a customer. Consultant B does not understand the difference between asking for an appointment and closing a sale after an estimate has been given. This one was a little sneaky if you did not read carefully. ASE test questions are intended to be read carefully. Many candidates miss questions because they did not read and comprehend the question before they chose their answer.

Question #37
Answer A is correct. Technicians use pH and freeze point information to determine the condition/mix of the coolant during service.
Answer B is wrong. Often sold with cooling system service, it is not part of the service unless there is leakage.
Answer C is wrong. Heater cores are not replaced during routine cooling system service.
Answer D is wrong. Adding refrigerant is part of an air conditioning system service, not a cooling system service. Word choice is important in discussion. Coolant and refrigerant are often confused in a pressure test.

Question #38
Answer A is wrong. Service Consultant B is also correct.
Answer B is wrong. Service Consultant A is also correct.
Answer C is correct. Both Service Consultants are correct. There are other sources for the maintenance schedule, but the main source for the vehicle owner is the owner's manual. Maintenance schedule choice is based on environmental and use parameters. Regular use verses severe condition use.
Answer D is wrong. Both service consultants are correct.

Question #39
Answer A is wrong. Service Consultant B is also correct.
Answer B is wrong. Service Consultant A is also correct.
Answer C is correct. Both Service Consultants are correct. Service Consultant A is correct. Creating expectations before you know the real answers sets the whole service department up for failure because you cannot know what is going to come up in diagnosis. This statement would exclude maintenance operations that you schedule by the hour and for which you can give accurate completion times. Don't make promises that are unrealistic. Get the answers first by diagnosing or inspecting the vehicle.
Answer D is wrong. Both Service Consultants are correct.

Question #40
Answer A is correct. A recall is mailed to a vehicle owner not a technical service bulletin.
Answer B is wrong. A Technical Service Bulletins (TSB) may explain a redesign of a component.
Answer C is wrong. A TSB may revise a shop manual procedure.
Answer D is wrong. A TSB may explain a pattern failure in a vehicle or group of vehicles.

Question #41
Answer A is wrong. Most contracts are tied to a mileage limit, so it would be necessary to have this information.
Answer B is wrong. The VIN is the unique identifying item for a vehicle and is almost always used to identify a service contract for the vehicle.
Answer C is wrong. Most contracts are also tied to a time limit based on the day the vehicle was put into service (purchased), so it would be necessary to have this information.
Answer D is correct. Service contracts and warranties are based on the day the vehicle was put in service, not the day it was built.

Question #42
Answer A is wrong. Extended hours may be a nice convenience for a customer, but only a very small number would do business with you solely on that.
Answer B is wrong. Appearance is very important, but is really not the best answer here and really part of creating trust.
Answer C is correct. When a customer trusts you, he or she is more likely to do business with you.
Answer D is wrong. It might be that some customers will do business with a shop based on discount prices, but they tend to have no loyalty to the shop and will go elsewhere if the deal is better. Statistically, price is several numbers below trust and how the customer is treated. If there are issues with the facility or the repair, the (low) price is quickly forgotten.

Question #43
Answer A is wrong. Dates stamped on valve covers could be the day the engine was built or even the day the valve cover was made.
Answer B is correct. There are dozens more but this is the only correct option in the list of choices provided.
Answer C is wrong. The emissions sticker is often on the radiator support or under the hood.
Answer D is wrong. Some manufacturers put their calibration code on the gas door, but usually it is just instructions about opening the cap and fueling the vehicle.

Question #44
Answer A is wrong. A throttle sensor is an engine management input.
Answer B is wrong. A PCM is a component of an engine management system.
Answer C is wrong. Wheel speed sensor is part of the anti-lock brake system.
Answer D is correct. The clock spring or spiral cable keeps constant electrical contact between the air bag module in the steering wheel and the air bag module when the steering wheel turns.

Question #45
Answer A is wrong. Service Consultant B is also correct.
Answer B is wrong. Service Consultant A is also correct.
Answer C is correct. Both Service Consultants are correct. While you hate to have a vehicle not get done as promised, it is much better to know so that you can offer the customer options before he or she is standing at your service counter to pick up the vehicle. Making good use of your time is always important.
Answer D is wrong. Both Service Consultants are correct.

Question #46
Answer A is wrong. TSBs are corrections to engineering mistakes; many may not be of strong importance.
Answer B is wrong. Very few campaigns are publicized by the manufacturer past their own dealer network. A recent new law change by the National Highway Traffic Safety Association will change the status of all campaigns to recall.
Answer C is wrong. There is no need to announce by letter the end of a warranty period.
Answer D is correct. The recall has a mandatory owner letter since they are usually safety, use, or value related repairs.

Question #47
Answer A is wrong. Service Consultant A is correct, but new information from Technician B would be helpful.
Answer B is wrong. Service Consultant B is correct, but a history of repair from Technician A is important as well.
Answer C is correct. Both Service Consultants are correct. Some customers might give you the impression that your facility has serviced the vehicle before, but you may find that there is not really a prior repair at your facility but rather that another facility has looked at it before. In Consultant B's statement, many times the customer may give you new information that will help to solve the problem.
Answer D is wrong. Both Service Consultants are correct. A check of history combined with new information may result in a fix.

Answers to the Test Questions for the Additional Test Questions Section 6

1. C
2. D
3. B
4. B
5. D
6. A
7. B
8. D
9. C
10. C
11. B
12. A
13. B
14. B
15. B
16. C
17. D
18. B
19. A
20. B
21. B
22. B
23. C
24. A
25. B
26. D
27. A
28. D

Explanations to the Answers for the Additional Test Questions Section 6

Question #1
Answer A is wrong. Service Consultant B is also correct.
Answer B is wrong. Service Consultant A is also correct.
Answer C is correct. Both Service Consultants are correct. This is an example of the difference between features and benefits. If you are still unclear on this area, refer to the information in task C.5 in Chapter 4.
Answer D is wrong. Both Service Consultants are correct.

Question #2
Answer A is wrong. A converter requires heat to work; closest to the engine is best.
Answer B is wrong. A converter requires heat to work; closest to the engine is best.
Answer C is wrong. The air pump is not part of the exchange system.
Answer D is correct. There are oxygen sensors behind the converters in OBD-II vehicles that monitor catalyst efficiency too. The idea here is that the computer checks exhaust gases before the converter and that the converter is mounted very close to the front of the vehicle to get them up to around 600 degrees as quickly as possible.

Question #3
Answer A is wrong. Seat belts are primary restraints.
Answer B is correct. Some manufacturers also call the system SIR or supplemental inflatable restraints. Many cars for years had only primary restraints; supplements increase safety and have become commonly included on modern vehicles.
Answer C is wrong. It controls the top of the engine.
Answer D is wrong. It is another primary restraint system.

Question #4
Answer A is wrong. The PCV valve reintroduces crank case gases to be burned in the engine.
Answer B is correct. A fuel injector is a component of the fuel system.
Answer C is wrong. The air pump delivers air into the exhaust stream to help burn leftover gases and to support catalytic converter efficiency.
Answer D is wrong. The fuel cap controls the fuel vapor and keeps the system closed to control hydrocarbon emissions.

Question #5
Answer A is wrong. Thrust angle is the relationship between the overall toe direction of the rear wheels to the overall toe direction of the centerline of the chassis.
Answer B is wrong. When you look at the wheel from the front, the amount the tire leans in or out at the top is called camber. When the tire leans out, it is called positive camber and, when it leans in at the top, it is called negative camber.
Answer C is wrong. Caster is an imaginary line drawn through the ball joints or from the top of the strut mount to the lower ball joint.
Answer D is correct. Toe is the direction the front of the wheel points when going down the road. If you stand up and point your toes toward each other, that is what is known as toe in. If you point them away from each other, that is known as toe out. The measurement of distance front to rear if the same, would be an "O" toe.

Question #6
Answer A is correct. Only Service Consultant A is correct. Different customers require different levels of explanation. If a customer wants lots of details, that does not automatically mean he or she distrusts you. It usually means the customer takes a very active part in maintaining the vehicle and wants to understand the process. These can be your very best customers. This is a great chance for you to show off your expertise and product and repair knowledge.
Answer B is wrong. This has nothing to do with trust.
Answer C is wrong. Only Service Consultant A is correct. Service Consultant B is wrong. Questions do not mean lack of trust, yet great answers can create trust.
Answer D is wrong. Only Service Consultant A is correct. Questions are expected at vehicle pickup.

Question #7
Answer A is wrong. A customer that only drives 500 miles should service the vehicle in the months driven, since they do not meet the miles driven the vehicle still should be serviced on its regular monthly interval schedule.
Answer B is correct. Only Service Consultant B is correct. If a customer meets the mileage first before the month of a scheduled service the vehicle should be serviced then and not wait.
Answer C is wrong. Only Service Consultant B is correct.
Answer D is wrong. Only Service Consultant B is correct.

Question #8
Answer A is wrong. We are looking for the best answer and D is correct. Wrong information can create wrong repairs.
Answer B is wrong. D is the best answer.
Answer C is wrong. Since the technician has not inspected the vehicle yet, you cannot provide an accurate estimate for repairs. It is very important before you let the customer go, or take them home, to make sure you have all of their information correct, so answer C is definitely wrong. You should never offer an estimate for repairs before you let your technician inspect the vehicle and confirm the customer's requests and concerns. You may have some menu priced items that you can offer prices for and that would be the exception to the rule. The other answers are things you would do but not until you have customer information confirmed.
Answer D is correct. Here we are looking for the best answer. The next move is to confirm the accuracy of all of the information you have collected before the customer leaves. Wrong information can create problems for the entire repair and transaction.

Question #9
Answer A is wrong. Service Consultant B is also correct.
Answer B is wrong. Service Consultant A is also correct.
Answer C is correct. Both Service Consultants are correct. Alternate transportation is any way the customer can get where they need to go. It can be the deal breaker for some customers who depend on their vehicle for transportation.
Answer D is wrong. Both Service Consultants are correct.

Question #10
Answer A is wrong. This describes toe out.
Answer B is wrong. This describes positive camber.
Answer C is correct. This is negative camber. Likely has a smooth inside tire wear pattern.
Answer D is wrong. This describes positive caster.

Question #11
Answer A is wrong. This is a component of the fuel system.
Answer B is correct. Just checking to see if you know what DIS stands for. Clearly, if an ignition system is distributorless, it will not have a distributor. DIS is the industry standard anagram for distributorless ignition system or direct ignition system.
Answer C is wrong. Part of a DIS system provides information to start the sequence of ignition system.
Answer D is wrong. These systems do not have an ignition system; they use high compression to ignite the fuel mixture.

Question #12
Answer A is correct. The distributorless system fires the plugs during both the power and exhaust stroke on most applications. This is usually due to the fact that the coils are shared by two opposing cylinders. This is often referred to as a waste spark ignition system.
Answer B is wrong. Plugs only fire on the power stroke in this ignition system.
Answer C is wrong. Plugs only fire on the power stroke in this ignition system. They fire the power stroke by heat created by high compression of the fuel/air mixture.
Answer D is wrong. Diesels do not have spark plugs.

Question #13
Answer A is wrong. No interference means no risk of piston-to-valve damage.
Answer B is correct. Only Service Consultant B is correct. Read the question carefully. A non-interference engine is one that will not have valve to piston contact if the timing belt breaks. A is wrong. B is correct. The camshafts are driven by the timing belt which is connected to the crankshaft.
Answer C is wrong. Only Service Consultant B is correct.
Answer D is wrong. Only Service Consultant B is correct.

Question #14
Answer A is wrong. The Service Consultant should discuss the issue with the customer, not estimate a tune-up.
Answer B is correct. The customer may have no idea but, by asking them, you have told them that you are conscientious and would not want to replace a component that is not needed. It is a really good idea to check your own history before you ask the customer to make sure the component was not replaced by your facility.
Answer C is wrong. It may look original, but has been replaced. Always ask before suggesting.
Answer D is wrong. Consult with the customer first. It may be an opportunity to sell a proper main service.

Question #15
Answer A is wrong. The filter is located inside a pan and may be serviceable or not.
Answer B is correct. Master cylinders, in newer transmissions referred to as the strainer, do not have any kind of filter and are sealed to containments.
Answer C is wrong. Filters are used to capture any small particles that could circulate and cause damage to the compressor, or restrict refrigerant flow causing cooling performance concerns.
Answer D is wrong. Many late-model vehicles use a filter in the heating/air conditioning system to filter out particulates from the incoming air. Much like a furnace filter, cabin smoke and/or misc. airbornes are filtered and trapped.

Question #16
Answer A is wrong. It connects the crankshaft to the transmission.
Answer B is wrong. Seals oil from leaking at the crankshaft rear.
Answer C is correct. If you are unclear about this return, to Section 4 and review the components in the exploded view and their descriptions.
Answer D is wrong. It is bolted to the cylinder block.

Question #17
Answer A is wrong. The service consultant should prioritize the complaints in order of importance.
Answer B is wrong. This order may cause the items that may be more important left not done.
Answer C is wrong. This would cause the customer to think you only want the money and they would only go elsewhere for service.
Answer D is correct. This is the correct order to prioritize and repair the most important items first.

Question #18
Answer A is wrong. A used replacement rim is not recommended since you do not know if this rim is defective and you could have issues in the future.
Answer B is correct. To recommend replacing all the rims to match is the correct response since it leaves the decision up to the customer and keeps safety in mind.
Answer C is wrong. This response would only cause a customer to be dissatisfied and could cause a safety concern.
Answer D is wrong. Even if this could be true it is not correct to use scare tactics to try a get a customer to buy something.

Question #19
Answer A is correct. When a customer is upset, the best thing you can do is let them vent. If they are too loud or you feel embarrassed, you might ask them to follow you to a private place where you can listen and help resolve the issue completely. This isolation can help control a temper until a remedy is reached.
Answer B is wrong. They already are upset to be there. This will make matters worse.
Answer C is wrong. It is best to listen. Don't tell them they are wrong too early in the discussion.
Answer D is wrong. Discounts don't solve problems. They only eat away at the shop's profitability.

Question #20
Answer A is wrong. The planetary is a multiple ratio gear set used in automatic transmissions to provide different gear ratios in a compact package.
Answer B is correct. The pressure plate is a component of the clutch in a manual transmission. All of the other items are components of an automatic transmission.
Answer C is wrong. The torque converter is a hydraulic coupling device that connects the engine to the transmission. Most late model automatics have a electric locking clutch inside the converter that makes a direct connection between the engine and the transmission during light load cruising. This drops rpm and helps with gas mileage.
Answer D is wrong. The valve body is the hydraulic control unit. Newer electronic transmissions have solenoids within the valve body that are run by the PCM.

Question #21
Answer A is wrong. Air conditioning uses refrigerant, not anti-freeze.
Answer B is correct. Only Service Consultant B is correct. The air conditioning system is a separate system and does not use antifreeze. There are many types of antifreeze and most do not mix. It is very important to identify the type that belongs in the vehicle. Color is a common method of identifying antifreeze types.
Answer C is wrong. Only Service Consultant B is correct.
Answer D is wrong. Only Service Consultant B is correct.

Question #22
Answer A is wrong. The transmission is not related to starting.
Answer B is correct. The starter drive gear engages the flywheel ring gear to rotate the engine for starting.
Answer C is wrong. The crank does turn during start up but the starter turns it via the flywheel and its ring gear.
Answer D is wrong. The battery is a storage device and does not engage or disengage; it just provides the electrical power to do so.

Question #23
Answer A is wrong. The service consultant should not offer a repair if no problem exists.
Answer B is wrong. The manufacture would only refer the customer back to there dealer and the customer would not like the bouncing back and forth.
Answer C is correct. A customer that just bought a car that has a (CVT) continuously variable transmission may not be aware that these are different transmissions and they will not feel the shift points like a traditional automatic transmission through which you will feel each shift.
Answer D is wrong. If the customer is not made aware that the complaint is not a problem but a normal operation, then when would the customer know they are really having a problem?

Question #24
Answer A is correct. CV axles pass through or bolt to the wheel hub and connect the differential or transaxle to the wheels.
Answer B is wrong. Tires are on the wheels and the wheels bolt to the wheel hub assembly.
Answer C is wrong. The engine is not connected to the engine except in some cases where the engine provides a mounting point for a bracket or support.
Answer D is wrong. The flywheel connects the engine to the transmission.

Question #25
Answer A is wrong. You would be wasting your time here. Most customers will not even know what their VIN is, let alone be able to dictate it to you.
Answer B is correct. Trainers and customer service people all agree that the first thing to do is find out the name of the person you are talking to. They will feel more comfortable throughout the transaction.
Answer C is wrong. The prime item is the reason the customer was triggered to bring the vehicle in. It is important to find out what it is but it should really rank second or third in your list of items to collect.
Answer D is wrong. Most people do not know their license number.

Question #26
Answer A is wrong. To speak first can be a mistake; you must review the estimate.
Answer B is wrong. Don't discuss too many variables over the phone.
Answer C is wrong. You must not solve problems with discounts.
Answer D is correct. Sometimes customers play repair shops against each other like little kids do with their parents. We must not take the bait and bad mouth another shop. It does not raise us up above them. Kindly offer the customer an appointment to confirm the diagnosis for them, and tell them what kind of charge you anticipate for confirmation.

Question #27
Answer A is correct. This is the first component of the hydraulic portion of the brake system. When the brake pedal is depressed, it generates hydraulic pressure to apply the brake calipers and wheel cylinders.
Answer B is wrong. The system has many components but none are referred to as the master cylinder.
Answer C is wrong. The system has many components but none are referred to as the master cylinder.
Answer D is wrong. The system has many components but none are referred to as the master cylinder.

Question #28
Answer A is wrong. This is a phone call. Items must be verified.
Answer B is wrong. Do this when the customer is present. This is a phone call.
Answer C is wrong. Never ask the customer to diagnose.
Answer D is correct. The service consultant is charged with taking the customer's information and organizing it so that the technician can methodically approach each symptom or service request without visiting with you several times to see what was meant. A test drive to verify all complaints may be helpful.

Glossary

Accumulator Part of an air conditioning system which contains a desiccant that absorbs moisture from the refrigerant.

Air filter Part of an engine intake system which cleans dirt and dust from the air before it enters the engine.

Air Injection Reaction (AIR) An emissions control system which pumps air into the exhaust system to burn hydrocarbons, carbon monoxide, and oxides of nitrogen coming from the engine.

Air pump Part of the Air Injection Reaction (AIR) system which forces extra air to mix with the exhaust gases. This action causes the continued burning of any hydrocarbons and carbon monoxide remaining in the exhaust.

Alternating current (AC) An electrical current which flows alternately in two directions, forward and backward. It is produced by some form of mechanical device or motion, such as an alternator. Alternating current cannot be used to charge a battery. It must first be converted (rectified) to direct current for battery charging.

Alternator The electrical device driven by the engine which recharges the battery. It produces alternating current using rotating field coils inside stationary stator windings.

Antilock brakes A brake system which operates by pulsing the pressure to the wheel and caliper cylinders to prevent the lockup of any one wheel, thus preventing a skid.

Automatic transmission A transmission which automatically selects the correct gear ratio needed for the driving conditions. The driver has only to select the direction of travel desired.

Ball joint Part of the steering system which connects the spindle to the control arm. It allows the steering knuckle to turn right and left as well as permitting the control arm to move up and down.

Band A flexible flat piece inside an automatic transmission which holds parts of the gearset to create the correct gear ratio.

Battery A device within the electrical system which stores voltage until it is needed for vehicle operation.

Bearing A term used for ball bearing; an antifriction device having an inner and outer race with one or more rows of hardened steel balls between them.

Blend door Part of the air conditioning/ventilation system which controls the ratio of incoming heated air and fresh air. This ratio is adjusted to control the temperature of the passenger compartment.

Blowby The unburned fuel and combustion byproducts which leak past the piston rings and into the crankcase.

Brake pads The friction elements of a caliper-type brake system. They are usually forced toward the rotor by hydraulic pressure.

Brake shoes The friction elements of a drum-type brake system. They are usually expanded outward by hydraulic pressure to contact the inside diameter of the brake drum.

Bushing A smooth cylinder used to reduce friction and to guide the motion of rotating parts.

Camshaft The shaft in an engine which causes the valves to open and close at the correct times.

Carbon dioxide A colorless, odorless, incombustible gas which is exhaled by humans and required by plant life. In the automotive emissions system, carbon monoxide is mixed with air (oxygen) to form harmless carbon dioxide.

Carbon monoxide A colorless, odorless, toxic gas formed by internal combustion engines.

Carburetor A device used on some engines to mix the fuel and air in the correct ratio for efficient combustion.

Catalytic converter A metal canister mounted in the exhaust system containing metals that convert harmful exhaust gases into safer gases. A catalytic converter speeds up the reaction but is not consumed in the chemical reaction.

Celsius The metric unit of temperature measurement.

Check valve A valve which allows something to move or flow in only one direction.

Circuit breaker An electrical device which protects the circuit by interrupting the current flow when it exceeds the rated capacity. It may be reset manually or automatically.

Closed loop An operating state in a computer-controlled engine in which the computer is controlling engine operation based upon information created by the sensors.

Closing a sale Getting a commitment to buy from a customer.

Clutch (1) The part of a driveline system which is used to interrupt the power flow between the engine and the transmission. (2) A part mounted on the air conditioning (AC) compressor used for temperature control of the AC system.

Clutch-release lever The part of a manual transmission system which operates the clutch-release (throwout) bearing.

Coil spring A spring which is wound into a spiral shape. Coil springs are commonly used on automotive suspension systems.

Combustion chamber The area inside the cylinder head and block where the burning of the fuel takes place.

Compression The act of forcing something together. In an automotive engine, the air and fuel are compressed in the cylinders to create a larger explosion and thus more power upon ignition.

Compressor The part of the air conditioning system which compresses the refrigerant vapor and pumps the refrigerant.

Condenser The part of the air conditioning system which cools the hot vapor and converts it to a liquid. The condenser is usually mounted in front of or on top of a vehicle for better airflow.

Constant velocity (CV) joint Part of the drivetrain which allows for changes in the angle of a driveshaft or half shaft.

Control arm Suspension parts which control spring action and the direction of travel of the axle as it reacts to driving conditions.

Coolant A fluid used for cooling, usually consisting of a blend of water and antifreeze.

Countershaft A shaft used in transmissions to transfer the motion from the input shaft to the output shaft.

Crankcase The lower part of an engine.

Crankshaft The shaft in an engine which converts the reciprocating piston motion into rotary motion for the driven device.

Crossover pipe The pipe which connects the two exhaust pipes of a dual exhaust system.

Cubic inch The volume equal to a cube with one-inch sides. The term is commonly used to describe engine displacement.

Customer relations A description of how a salesperson interacts with the customer.

Cylinder head The removable part covering the top of the engine cylinders. It seals the combustion chamber and usually contains the valves and spark plugs.

Diameter The distance straight across a circular figure; the largest measurement which can be taken across a circular object.

Differential The set of gears which transmits power from the driveshaft to the wheels and allows the drive wheels to turn at different speeds for cornering.

Direct current (DC) An electrical current which flows in only one direction. It is usually created from a chemical source, such as a battery, and can be used to recharge a battery.

Distributor The part of the ignition system which directs the secondary voltage (spark) to the correct spark plugs and usually drives the oil pump.

Distributorless Describes an ignition system that uses crank and or cam position sensors to control secondary voltage to the spark plugs.

Drag link The rod which connects the steering box to the steering knuckle on a straight axle front.

Driveshaft The shaft which connects the transmission to the differential. in rear-wheel and 4WD applications

Engine block The main body of an engine. The block contains the cylinders and carries the accessories.

Engine coolant The solution used in an engine to carry heat away from the cylinders. It is usually a 50/50 mixture of ethylene glycol and water.

Evaporator The part of an air conditioning system which is mounted inside the passenger compartment. It contains low-pressure refrigerant liquid and vapor to remove heat from the air forced through and around it.

Exhaust Gas Recirculation (EGR) An emissions device which meters some exhaust gas into the intake manifold to reduce the combustion chamber temperature and to prevent the formation of oxides of nitrogen.

Exhaust manifold The part of the exhaust system which bolts directly to the cylinder heads. Exhaust gases leaving the cylinder are routed through the cylinder head to the exhaust manifold.

Fahrenheit The temperature unit used in the English system of measurement.

Fuel injector The fuel system component which sprays fuel into the intake manifold on vehicles not equipped with a carburetor.

Fuel pump The fuel system component which moves the fuel from the tank to the carburetor or the fuel-injection unit.

Fuse An electrical device which protects the circuit by interrupting the current flow when the flow exceeds the fuse's rated capacity. Fuses are constructed of a conductor encased in plastic or glass and must be replaced when blown.

Fusible link An electrical device which protects the circuit by interrupting the current flow when the flow exceeds the rated capacity. A fusible link is constructed of a wire surrounded by a special insulation and must be replaced when blown.

Gross profit The selling price of an item or service less the cost.

Half shaft One of two shafts which connect a transaxle to the drive wheels or connects the differential to the hub in a 4WD or independent rear-wheel drive application.

Hydrocarbon An organic compound made up of hydrogen and carbon. Most automotive fuels and lubricants contain hydrocarbons, a common source of pollution. As an automotive term, it refers to unburned fuel in the exhaust.

Hydrogen A colorless, highly flammable gas which is the most common in the universe. It is used in the production of methanol, an automotive fuel and fuel additive.

Idle A condition in which the engine is running at a low speed.

Idler arm A part of the steering system which connects the center link to the frame.

Ignition coil A part of the ignition system which produces the electricity needed to create the spark for the spark plugs.

Ignition system The part of the automotive electrical system which creates and delivers the high-voltage spark to the spark plugs.

Input shaft The shaft which carries torque into the transmission.

Intake manifold The part of the intake system which bolts directly to the cylinder heads. The air or air/fuel mixture must pass through the intake manifold before entering the cylinder head and combustion chamber.

Intake valve The valve in the intake port which opens to allow the air/fuel mixture to enter the combustion chamber.

Intercooler A part of the intake system used with a turbocharger to cool the air entering the intake manifold.

Integrated circuit A miniaturized electrical component-consisting of diodes, transistors, resistors, and capacitors-used in electronic circuits.

Journal The part of a shaft which is supported by and comes in contact with a bearing.

Lifter A component of the automotive valve train which converts the rotary motion of a camshaft lobe into the reciprocating motion needed to open and close the valves.

Limited-slip differential A differential which uses internal clutch plates to limit the slip and speed difference between the drive wheels. This limiting improves traction on slick surfaces and helps eliminate wheel spinning.

Liter A unit of volume used in the metric system of measurement. A liter is slightly greater than a quart in the English system.

MacPherson strut A suspension system component which combines a lower lateral link with a vertical strut to combine the features of a spindle and a shock absorber.

Main bearing caps The caps which secure the crankshaft to the block.

Material Safety Data Sheet (MSDS) A prepared printed sheet which accompanies chemicals and contains information regarding the proper application along with the safety and environmental concerns.

Microprocessor A small computer-like device used to process signals from various circuits to achieve a control system that will adapt to operating changes.

Muffler The part of the exhaust system which is used to reduce exhaust noise.

Net price The cost of an item to a particular purchaser, such as dealer net, jobber net, or user net.

Net profit The merchant's profit after deducting costs of merchandise and all expenses involved in operating the business.

OE An abbreviation for original equipment.

OEM An abbreviation for original equipment manufacturer.

Oil filter A device on the engine for removing dirt, carbon, and other impurities from the lubricating oil.

Open loop An operating state in a computer-controlled engine in which the computer is controlling engine operation based upon a predetermined program. It is usually in effect until the engine sensors signal that the engine has reached operating temperature.

Orifice tube A part of an air conditioning system which causes the drop in refrigerant pressure that results in the change in temperature necessary for cooling.

OSHA An abbreviation for Occupational Safety and Health Administration. All companies with a specified minimum number of employees must comply with its safety regulations.

Overhead The costs of operating a business not including purchases of merchandise.

Overhead cam An engine valvetrain system which has the camshaft positioned on top of the cylinder head.

Oxygen A gas making up approximately 18 percent of the air in the atmosphere and required for the combustion process to take place in an engine.

Oxygen sensor A computer system sensor which monitors the oxygen content in the exhaust gas. The signal from the oxygen sensor tells the computer whether a rich or lean condition exists in the air/fuel ratio entering the engine.

Parking brake A mechanical brake on the vehicle used for parking or emergency stopping situations.

Pilot bearing A bearing mounted in the center of the crankshaft which supports the end of the transmission input shaft.

Pinion yoke The yoke mounted on the end of the pinion gear of the differential. The pinion yoke transfers torque from the driveshaft to the pinion gear.

Port fuel injection An automotive fuel delivery system which has a fuel injector for each cylinder positioned in the intake manifold at the base of the intake valve.

Positive crankcase ventilation (PCV) An emissions system which draws hydrocarbons from the engine crankcase and routes them through the intake manifold to be burned in the engine.

Professional One who performs a specialized service as a means of employment, such as an ASE-certified automotive technician.

Profit The amount received for goods or services above the amount of expenses.

Pushrod A valvetrain component used in engines to connect the lifter to the rocker arm.

R-12 The trade name for a refrigerant commonly used in automotive air conditioning systems. It has been phased out because of its hazard to the earth's ozone layer. R-12 is being replaced by R-134A as an automotive refrigerant.

R-134A The trade name for a refrigerant currently used in automotive air conditioning systems. It is the replacement of choice for R-12.

Rack and pinion A steering system which uses a horizontal rack with gear teeth and a pinion gear attached to the end of a steering shaft.

Radiator The device mounted at the front of the vehicle which is used to cool the engine. The hot coolant flows through the radiator. The radiator fins contain the coolant, and air flowing past the fins will remove heat.

Relay An electrical device which allows the remote control of a switch. It normally allows the control of a large current with a much smaller current.

Remanufactured part A part which has been reconditioned to original standards.

Resonator A type of secondary muffler.

Right-hand rule Store layout designed to permit customers to move naturally to the right on entering.

Rod bearing caps Rod bearing caps hold the rod to the crankshaft.

Rotor A part of the brake system which turns with the wheel spindle and is clamped by the brake pads for braking action. It is a term that also refers to the rotating part of an alternator that contains the field windings.

SAE An abbreviation for Society of Automotive Engineers, which sets the standards for many products.

Selling up Selling a customer a better quality item when a lower priced item obviously will not do the job he or she expects of it.

Service bulletin A bulletin which provides supplemental and update information for service manuals.

Service manual A manual which contains the diagnosis and repair procedures for vehicle systems.

Servo The hydraulic piston which applies the bands found in an automatic transmission.

Spark plug The device which ignites the air/fuel mixture in the combustion chamber.

Stud A fastening device which resembles a bolt. It has threads at each end and no head.

Thermostat (1) The device which controls the water flow through the radiator. It prevents water from flowing through the radiator until the engine is at operating temperature. (2) The device that controls the temperature of an air conditioning system.

Throttle body A fuel-injection system which places the fuel injectors in a housing at the location previously used by the carburetor.

Throwout bearing A part of the drivetrain used with manual transmissions. It presses in the pressure plate fingers to release the clutch. It is also known as the clutch-release bearing and is mounted on the clutch-release lever.

Thrust angle The alignment of the rear axle to the vehicle.

Tie-rod end The movable joint which connects the tie rods to the steering knuckle. Wear in the tie-rod ends will cause excessive tire wear.

Timing belt A belt which connects the camshaft and crankshaft, synchronizing their rotation. Timing belts are normally used with overhead cam engines.

Timing chain A chain which connects the camshaft and the crankshaft, synchronizing their rotation.

Timing gears The gears mounted on the end of the camshaft and crankshaft. They mesh together and synchronize the rotation of the camshaft and crankshaft.

Torque converter The device located between the engine and the automatic transmission. It will slip and allow the engine crankshaft and automatic transmission input shaft to rotate at different speeds.

Transaxle A combination transmission and differential.

Transfer case A gearbox which connects the front and rear drivelines of a four-wheel drive vehicle.

Transmission A set of gears which can change ratios to meet the various driving needs of the vehicle.

Transverse An orientation which refers to the engine and transaxle being mounted crosswise in the vehicle.

Turbocharger A device mounted on the engine which forces air into the intake manifold. Turbochargers are driven by exhaust gas from the engine and serve to boost power and performance.

Universal joint (U-joint) A flexible coupling which connects the driveshaft to the pinion yoke of the differential.

Vehicle identification number (VIN) A unique number assigned to a vehicle for identification purposes. The VIN number can be decoded for information regarding year of manufacture, manufacturer, body style, engine size, carrying capacity, and other information.

Voltage regulator The charging system device which sets the maximum charging voltage produced by the alternator.

Warranty A printed document stating that a product or service will provide satisfactory service for a given period of time or the buyer will be entitled to a settlement according to the terms of the warranty.

Water pump The pump located at the front of the engine which circulates engine coolant throughout the cooling system.

Wheel bearings The bearings located between the wheel hubs and spindle or axle housing and the axle.

Notes

Notes

Notes

Notes

Notes

Notes

Notes

Notes

Notes

Notes